21 Advances in Polymer Science

Fortschritte der Hochpolymeren-Forschung

Edited by H.-J. CANTOW, Freiburg i. Br. · G. DALL'ASTA, Cesano Maderno
J. D. FERRY, Madison · H. FUJITA, Osaka · M. GORDON, Colchester
W. KERN, Mainz · G. NATTA, Milano · S. OKAMURA, Kyoto · C. G. OVER-
BERGER, Ann Arbor · T. SAEGUSA, Yoshida · G. V. SCHULZ, Mainz
W. P. SLICHTER, Murray Hill · A. J. STAVERMAN, Leiden · J. K. STILLE,
Iowa City

With 68 Figures

Springer-Verlag
Berlin Heidelberg GmbH 1976

Editors

ISBN 978-3-662-15851-7   ISBN 978-3-540-38238-6 (eBook)
DOI 10.1007/978-3-540-38238-6

Library of Congress Catalog Card Number 61-642

© by Springer-Verlag Berlin Heidelberg 1976
Originally published by Springer-Verlag Berlin Heidelberg New York in 1976
Softcover reprint of the hardcover 1st edition 1976

Contents

Contents

Poly(isobutylene-*co*-β- Pinene)
A New Sulfur Vulcanizable, Ozone Resistant Elastomer by Cationic Isomerization Copolymerization

I. Synthesis and Reactivity Studies

Joseph P. Kennedy and Tom Chou*

The Institute of Polymer Science, The University of Akron, Akron, Ohio 44325, U.S.A.

Table of Contents

* Tom Chou present address: Arlon Products, 19200 Laurel Park Road, Compton, California 90224, U.S.A.

I. The Problem: Objectives and Significance

The first objective of this research was to demonstrate experimentally the feasibility of synthesizing well defined statistical isobutylene-β-pinene copolymers. A further objective was the detailed characterization of the copolymers and ultimately, the elucidation of the structure-property relationship in this system.

Our plans also included the exploration of the fundamentals of this copolymerization in which one of the monomers, β-pinene, isomerizes prior to propagation. In this sense, this research concerned the investigation of an "isomerization copolymerization", a field that has not yet been explored. This phase of our work involved the elucidation of the effect of a variety of initiating systems (counter-ions), solvents, and temperatures on the reactivity ratios of these monomers, on the rate of polymerization, and on the molecular weights of the products.

Prior to commencing our experiments, we theorized that isobutylene and β-pinene should copolymerize by a cationic mechanism. This expectation was based on the following two facts:

1. There is a chance for copolymerization if the two component monomers can be homopolymerized by the same type of mechanism, i.e., cationic, anionic, radical, oxonium ion, etc. Both isobutylene and β-pinene are highly reactive cationic monomers and as such can readily be polymerized by the same cationic initiator systems, e.g., $AlCl_3$, BF_3, etc. to high polymers.

2. There is a good chance for copolymerization if the two monomers exhibit similar reactivities toward a common growing species. Both isobutylene and β-pinene exhibit high reactivities under electrophilic polymerization conditions. And, significantly, the propagating species that participates in these polymerizations are very similar:

$$R-\overset{\displaystyle /CH_3}{\underset{\displaystyle \backslash CH_3}{C\oplus}}$$

where R = $-CH_2-$ for isobutylene and R= ⟨ ⟩ for β-pinene. This interesting circumstance arises as a result of the two individual propagating mechanisms involved:

Isobutylene: ~~~$\oplus$ + $CH_2{=}\overset{/CH_3}{\underset{\backslash CH_3}{C}}$ ⟶ ~~~$CH_2{-}\overset{/CH_3}{\underset{\backslash CH_3}{C\oplus}}$

β-Pinene: ~~~$\oplus$ + $CH_2{=}$⟨⟩ ⟶ ~~~$CH_2{-}$⟨⟩ $\overset{}{\underset{\sigma}{\longrightarrow}}$ ~~~$CH_2{-}$⟨⟩$\overset{CH_3}{\underset{CH_3}{{-}C\oplus}}$

In the cationic polymerization of isobutylene, the propagating electrophile is the

$$-CH_2-\overset{\displaystyle /CH_3}{\underset{\displaystyle \backslash CH_3}{C\oplus}}$$

structure. Isobutylene is an "ideal" cationic monomer as it gives a tertiary growing cation that cannot rearrange during propagation.

The mechanistic details of the cationic propagation of β-pinene as shown by the above equation have been established by Roberts and Day (1) and subsequently by a series of independent investigators (2–5). According to the now generally accepted view, propagation cannot involve the initially formed tertiary cation

$-CH_2-$ because of excessive steric hindrance toward propagation. Although

the initial tertiary cation is unable to propagate, the β-pinene skeleton still provides a favorable mechanism for propagation by the opening of its highly strained 4-membered condensed ring system:

Indeed this relief of strain provides the necessary driving force for rapid polymerization. Thus exomethylene cyclohexane cannot be polymerized to high polymer (1, 6) as discussed in the framework of general polymerizabilities of cyclobutane derivates (7, 8).

Significantly, then, the propagating sites of isobutylene and β-pinene polymerizations are, for all practical purposes, identical. The crucial cross-propagation events must also have quite similar energy contents because not only the growing cation sites but also the sites of highest electron density (the double bond) of these monomers are very similar:

These structural-mechanistic considerations strongly suggested the feasibility of random copolymerization of isobutylene and β-pinene.

Homopolymers of isobutylene and β-pinene are, respectively, nonvulcanizable rubbers and brittle plastics. The projected structure of a random poly(isobutylene-co-β-pinene) is of interest, since depending on the relative composition of the copolymer structure e.g.,

a wide variety of physical properties may be anticipated:

1. With only small amounts (< 10%) of β-pinene units in the chain rubbery copolymers may be expected. These novel rubbers should be vulcanizable with sulfur

(six allylic hydrogens) and should be completely ozone resistant (internal "protected" double bond in the ring).

2. Intermediate amounts of β-pinene (20–60%) in the chain should shift the overall balance of properties toward "leathery" materials. As the proposed structure consists of

"soft" and "hard" units,

properties that are intermediate between rubbers and rigid plastics are expected in this range.

3. With large amounts (> 80%) of β-pinene in the chain tough, impact resistant plastics may be anticipated.

II. Background

While the cationic homopolymerization of isobutylene has been the subject of numerous books (9), recent book chapters (10, 11) and review articles (12, 13), and while the polymerization of β-pinene has also been studied by many authors (1–3, 5, 14, 15), there is a dearth of information on the copolymerization of isobutylene with β-pinene. A review of the homopolymerization literature of these two monomers falls outside the scope of this paper.

There are only two reports in the literature concerning the copolymerization of isobutylene and β-pinene and even these two are fundamentally contradictory.

On the one hand, in a U.S. Patent Ott (16) claims that these two monomers can be copolymerized, while on the other hand Finnish authors (17) maintain in a recent publication that under essentially similar conditions to those described in the patent, copolymerization does not occur. The method described in the patent consists of condensing the isobutylene gas in a reactor at −60° ± 10 °C, diluting it with ethyl chloride solvent, adding β-pinene (or other terpenes) and inducing the polymerization of the well agitated system by introducing gaseous BF_3. After a certain time (up to 90 mins) the reaction is quenched by the introduction of ethanol, the precipitating solids are washed with ethanol and dried. The products were characterized by visual examination (hard, brittle polymer, flexible, tacky, etc.) and by a "Hercules drop melting point" method in the range from 55° to 100 °C.

The Finnish authors (17) carried out their copolymerization under N_2 atmosphere in three-necked flasks. The isobutylene and β-pinene monomer charge in CH_2Cl_2 was stirred at −10°, −30° and −50 °C and the polymerizations were started by introducing BF_3 gas through a capillary tube. After thirty minutes the reactions were quenched with cold CH_3OH. For comparison purposes homopolymerizations of isobutylene and β-pinene have also been carried out. The products were characterized by fractionation, GPC, viscosities, VPO and IR spectroscopy. All the polymers had low molecular weight, 3,500–1,500. Chiefly on the basis of GPC coupled with molecular weight determinations the authors concluded that copolymerization of isobutylene with pinene has not taken place under any of their conditions.

This contradiction between the American and Finnish workers have not yet been resolved. In this paper it will be shown that Sivola and Harva (17) are in error, that they have committed an experimental error which led to incorrect conclusions and that isobutylene and β-pinene can readily be copolymerized to high molecular weight product of interesting physical properties.

III. Experimental

A. Materials

β-Pinene (98%, Glidden Chemical Company) was distilled on a Todd Precise Fractionation Assembly. Only fractions of 99 + % purity were used. Impurities (by gas chromatography): α-pinene < 0.4%, other impurities combined < 0.5%. Purified β-pinene was dried over Molecular Sieves (3 Å, powder) for twenty four hours before use. Gas chromatographic analysis of β-pinene before and after drying with Molecular Sieves gave identical results, indicating that this drying agent was inert and did not isomerize β-pinene.

Isobutylene, 99.0%; methyl chloride, 99.5%; and ethyl chloride, 99.7% (J. P. Baker Company) were dried by passing these gases through glass columns (120 x 5 cm ID) packed with porous barium oxide (Barium and Chemicals, Inc.) and Molecular Sieves (3 Å, powder).

All initiator solutions were prepared under nitrogen atmosphere in a stainless steel enclosure.

Neat $C_2H_5AlCl_2$ (Texas Alkyl Company) was vacuum distilled before use. $AlCl_3$ (Matheson Company) was refluxed several times for two hours in methyl chloride (5g $AlCl_3$/50ml CH_3Cl) using a distillation flask equipped with a condenser cooled with Dry Ice and pentane mixture. The yellow saturated $AlCl_3$ solution was decanted and discarded between refluxings until a crystal clear solution was obtained. The concentration of the (saturated) $AlCl_3$ solution has not been determined but is estimated to be ~0.5% (18). BF_3 (Matheson Company) was used as received without further purification. The BF_3 initiator solution was prepared by absorbing gaseous BF_3 in MeCl solvent at –90°. The exact amount of BF_3 absorbed has not been determined, thus, the concentration of the initiator solution is unknown.

B. Polymerizations

All reactions were carried out in a stainless steel enclosure under a nitrogen atmosphere. The moisture level in the atmosphere of the enclosure was kept below 100 ppm by passing a constant dry nitrogen gas stream through the dry box. The moisture level was measured by a Model W Water Analyzer (Manufacturing Engineering and Equipment Corporation).

Polymerizations were carried out in three necked flasks equipped with mechanical stirrer, and thermometer. The flask was placed in a thermostatically controlled bath filled with n-pentane and cooled by liquid nitrogen (*18*). Initiator solutions were precooled and added through a cold jacketed addition funnel to the stirred charges. Reactions were fast and exothermic.

The rate of initiator addition was slow (*e.g.* 2 ml/min of 0.5 volume % initiator solution) so as to minimize the temperature rise in the reaction flask. In practice, some temperature rise (2–10 °C) was always experienced. The polymer was insoluble in the charges and precipitated immediately. Precooled, dilute methanol was added to quench the reaction.

The liquid portion was decanted, the solid polymer collected, washed with methanol and then dissolved in n-pentane. The solution was coagulated by dropwise addition into methanol containing 0.1% antioxidant, 2,2-methylene-bis-(4-methyl-6-tertiarybutyl-phenol), or Cyanamide 2246. The polymers were again collected by decantation and then vacuum dried at 40 °C for 48 hours to constant weight. Gelation occurred on storage in the absence of antioxidant; in the presence of antioxidant gelation had not taken place even after one year of storage. Although gelled product was troublesome during our early characterization work, it was the first welcome indication for true copolymer formation.

C. Characterization

Intrinsic viscosities were measured using diisobutylene solvent at 20 °C and a Ubbelohde Viscometer.

The relationship between molecular weight and intrinsic viscosity for each copolymer composition was not established. An approximate value for $\overline{M}_v$ was obtained from Eq. (1) derived for polyisobutylene (*19*):

$$\ln \overline{M}_v = 12.48 + 1.565 \ln [\eta] \tag{1}$$

For copolymers containing large amounts of β-pinene, this approximation will give somewhat high $\overline{M}_v$ values because the chain-dimensions of β-pinene units are somewhat larger than those of isobutylene units (*19*).

Number average molecular weights were calculated from osmotic pressure, using a Mechrolab High Speed Osmometer 503 and toluene at 28 °C. The π/C vs. concentration plots gave straight lines in the concentration range from 0.02 to 0.10 *g/dl.*

Gel permeation chromatography was carried out on a Water Associates GPC Model 200, using seven matched columns: 2×10^3 Å to 5×10^3 Å, 5×10^3 Å to 1.5×10^4 Å, 5×10^3 Å to 1.5×10^4 Å, 1.5×10^4 Å to 5×10^4 Å, to 5×10^4 Å to 1.5×10^5 Å, 1.5×10^5 Å to 7×10^5 Å, 7×10^5 Å to 5×10^6 Å. The pressure was maintained at 140 psi, flow rate at 1 ml/min and temperature at 28 °C. The concentration of the samples was 0.25 *g/dl* in THF.

Copolymer composition was determined from PMR spectra obtained on a Varian T-60 Nuclear Magnetic Resonance Spectrometer using CCl_4 solutions and TMS as internal reference. The following equation was derived to calculate the relative amounts of isobutylene and β-pinene in the copolymer:

$$\beta\% = \frac{800\,A_0}{A + 7\,A_0} \tag{2}$$

where β% is mole percent β-pinene in the copolymer, A_0 is the area associated with olefinic protons at 5.3 ppm down field from the TMS signal, and A is the remaining area associated with all the aliphatic protons. Resolution of chemical shifts in the methyl region were obtained using a high resolution (300 Hz) Varian Model HR 300 NMR Spectrometer.

To check the accuracy of our method at low β-pinene values, we analyzed the PMR spectrum of an isobutylene-isoprene copolymer which according to commercial specifications contained 1.5 mole% unsaturation (This material, Butyl 268, has been obtained by the courtesy of the ENJAY Chemical Company). Using our spectroscopic method, we found 1.5 mole% unsaturation in this material.

D. Assessing the Possibility of Isomerization Preceding Polymerization

Since β-pinene readily isomerizes to α-pinene and other products in acidic media (20), it was important to examine the possibility of isomerization under our polymerization conditions. Thus a polymerization experiment has been carried out under conditions conducive for isomerization ([M] = 3.5 M, [EtAlCl$_2$] = 10^{-3} M, −70°, for 30 minutes). Conversion was about 6%. After quenching with methanol and work-up as described in Section B, the liquid charge was extracted with water (to remove methanol) and subsequently with n-pentane. The composition of the extract was analyzed by G.C. and compared with an original β-pinene sample. According to gas chromatographic analysis, β-pinene did not isomerize and the composition of the liquid charge was virtually identical before and after polymerization e.g., 98% β-pinene, 0.9% α-pinene and 1.1% unidentified compounds.

E. The Effect of Methanol Quenching Agent

Despite the extensive use of methanol as quenching agent in cationic polymerizations, little is known about the chemical reactions which occur during quenching. An observation related to this question is worth mentioning here.

To a mixture of 12 ml β-pinene and 3 ml methanol at −78°, 3 ml EtAlCl$_2$ was added. After about 10 minutes a viscous oil was obtained. The PMR spectra of the mixtures before and after reaction have been compared. It was found that the methyl proton singlet of methanol had shifted from 3.4δ to 3.1δ which indicates the presence of a methoxy proton in an ether linkage. According to this finding, the following reaction might have occured:

$$=\underset{}{\bigcirc}\text{Y} + CH_3OH \xrightarrow{\text{EtAlCl}_2} CH_3-\underset{}{\bigcirc}-\overset{\overset{\displaystyle CH_3}{|}}{\underset{\underset{\displaystyle CH_3}{|}}{C}}-O-CH_3 \qquad (3)$$

We did not uncover any literature about this ether and could not confirm the spectrum with an authentic sample.

IV. Results and Discussion

Table 1 is a compilation of relatively large scale copolymerization experiments carried out to collect material sufficient for characterization and physical property studies. Results of experiments conducted to elucidate the effects of reaction variables on reactivity ratios i.e., small scale runs (~20 ml), are listed separately later.

A. Characterization of Poly (isobutylene-co-β-Pinene)

A detailed structure characterization of isobutylene and β-pinene copolymers has been carried out including homogeneity studies (by GPC), quantitative composition and sequence analysis (by PMR and reactivity ratios) and molecular weight determinations (by osmometry and viscometry). Analysis of our data leads us to conclude that isobutylene and β-pinene can be readily copolymerized to reasonably high molecular weight materials and that the products are perfectly random, statistical copolymers showing no detectable tendency for "blockiness".

1. Gel Permeation Chromatography Studies

To reach a valid conclusion as to the overall composition of a copolymer, its homogeneity has to be established. In this sense we have investigated the homogeneity of our copolymers by GPC technique but did not seek quantitative molecular weight distribution information.

Beyond this, GPC characterization was of import since according to Sivola and Harva (17) when isobutylene and β-pinene in CH_2Cl_2 solvent are treated with $AlCl_3$ and/or BF_3 in the range $-10\,°C$ to $-50\,°C$, polymers exhibiting bimodal molecular weight distribution are produced. The low molecular weight fraction was found to consist essentially of poly(β-pinene) while the high molecular weight fraction was polyisobutylene. The presence of a possible copolymer was believed to be limited to the narrow range where the high and low molecular weight peaks partially overlapped. Indeed, on the basis of these findings, the Finnish authors concluded that isobutylene and β-pinene do not copolymerize.

Contrary to these results, we find that isobutylene and β-pinene readily copolymerize and that the copolymers are essentially homogeneous. The GPC trace of a representative product is shown in Fig. 1. This and other GPC traces show monomodal distribution, a near symmetrical shape with a slight skew toward low molecular weights. The $\overline{M}_v/\overline{M}_n$ values of our samples were in the 3–5 range.

A close examination of Sivola and Harva's paper (17) holds the key to the understanding of the discrepancy between these data and our findings. While we have terminated our runs at low conversions or conducted the copolymerizations under azeotropic conditions (i.e., by maintaining the monomer concentration or the charge

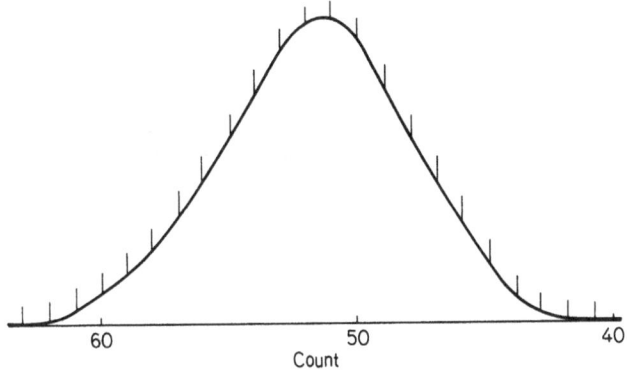

60   50   40
Count

Fig. 1. Gel permeation chromatogram of poly(isobutylene-co-β-pinene) prepared with EtAlCl$_2$ at −128°, conversion = 19%, β-pinene content = 10 mole %, M$_n$ = 3 x 10^4, sample 4 in Table 1

essentially unchanged, see later) the Finnish authors carried their experiments to high conversions (70–80%). Due to the large differences in the reactivity ratios between isobutylene and β-pinene in the temperature range −30 °C to −50 °C, the formation of heterogeneous mixtures including homopolymers and a variety of copolymers are expected at medium to high conversions. Had Sivola and Harva stopped their experiments at low conversions or had they conducted polymerizations at −100 °C and below i.e., in the azeotropic range (see later), they would have obtained homogeneous copolymer products as well.

2. PMR Analysis of Poly (isobutylene-co-β-Pinene)

The PMR spectra of a series of poly(isobutylene-co-β-pinene) have been recorded and analyzed. Figure 2 shows the PMR spectrum of a representative copolymer containing 24% β-pinene together with the spectra of respective homopolymers.

The resonance associated with the gem-dimethyl group in the copolymer appears as three partially resolved resonances at 1.1, 1.0 and 0.8 δ. These resonances have been assigned to "fully crowded", "half crowded", and "uncrowded" gem-dimethyl groups, respectively. The assignment of these resonances is based on an examination of the three spectra in Fig. 2. Thus, the spectrum of polyisobutylene, a molecule that contains only "fully crowded" gem-dimethyl groups i.e., gem-dimethyl groups flanked on both sides by other gem-dimethyl groups, exhibits a strong singlet at 1.1 δ. On the basis of this evidence we assigned the band at 1.1 δ in the

Table 1. Synthesis, characterization and physical properties of poly(isobutylene-co-β-pinene)

	1	2	3	4	5	6	7
Synthesis Details							
Total monomer concentration, isobutylene and β-pinene, mole/l of ethylchloride	3.4	3.4	3.4	3.2	3.2	2.4	2.1
Total Volume, l.	3.8	3.8	3.8	4.0	0.6	0.5	0.05
f_2 (mole% β-Pinene in feed)	3	3	3	10	20	40	70
EtAlCl$_2$ concentration, Vol.%	1	0.2	0.5	0.5	0.3	0.2	4
Volume of EtAlCl$_2$ solution used, ml. approx.	200	60	200	120	100	100	10
Temperature range, °C	−129→−114	−128→−120	−130→−120	−130→−127	−128→−123	−118→−115	−110→−108
Time, min.	20	40	46	138	28	22	10
Copolymer yield, g	490	38	151	142	8	4	2.5
Initiator efficiency, (g. polymer/mole EtAlCl$_2$)x10^{-3}	75	20	40	10	5	2	1
Conversion, on total monomers, wt.%	70	5	21	19	4	2.4	20
Characterization							
$[\eta]^{20}$, g/dl.	1.2	4.3	1.8	0.54	0.29	0.10	0.08
M_v x 10^{-3}	310	2,000	650	92	27	10	5
M_n x 10^{-3}		200	190	30			
F_2 (mole % β-Pinene in copolymer)	4	3	3	10	18	41	69
Physical appearance	White rubbery solid	White tough rubbery solid	White tough rubbery solid	White rubbery solid	Cohesive mass, retractable	Retractable with delayed response	Brittle powder

Physical Properties

Tg, °C	−65	−53	−28	+12
Sulfur vulcanizability	Vulcanizable	Vulcanized to strong rubber	Vulcanized to strong rubber	Vulcanized to strong rubber (and covulcanizable with Bd rubber)
Tensile strength, kg/cm²	35	Uncertain because of slippage; lower limit 211	240	148 (176 co-vulcanizate with PBd)
Modulus, 300%, kg/cm²	11	7.7	7.7	120 (176 co-vulcanizate with PBd)
Elongation, %	500	1,000	950	350 (300 co-vulcanizate with PBd)
Ozone resistance determined (see text for details)	Resistant		Resistant (stress-strain-unaffected)	

spectrum of the copolymer to fully crowded gem-dimethyl sequences, *i.e.*, isobutyl-
ene-isobutylene or ii for brevity, sequences. Poly(β-pinene), a molecule that contains "uncrowded" gem-dimethyl groups, shows a resonance band at 0.8δ. On the
basis of this, we assigned the band appearing at 0.8 δ in the spectrum of the copolymer to uncrowded gem-dimethyls indicating a β-pinene unit that is followed by
another β-pinene unit *i.e.*, $\beta\beta$ sequences. The central band at 1.0δ of the three partially unresolved bands in the spectrum of the copolymer is a new gem-dimethyl band

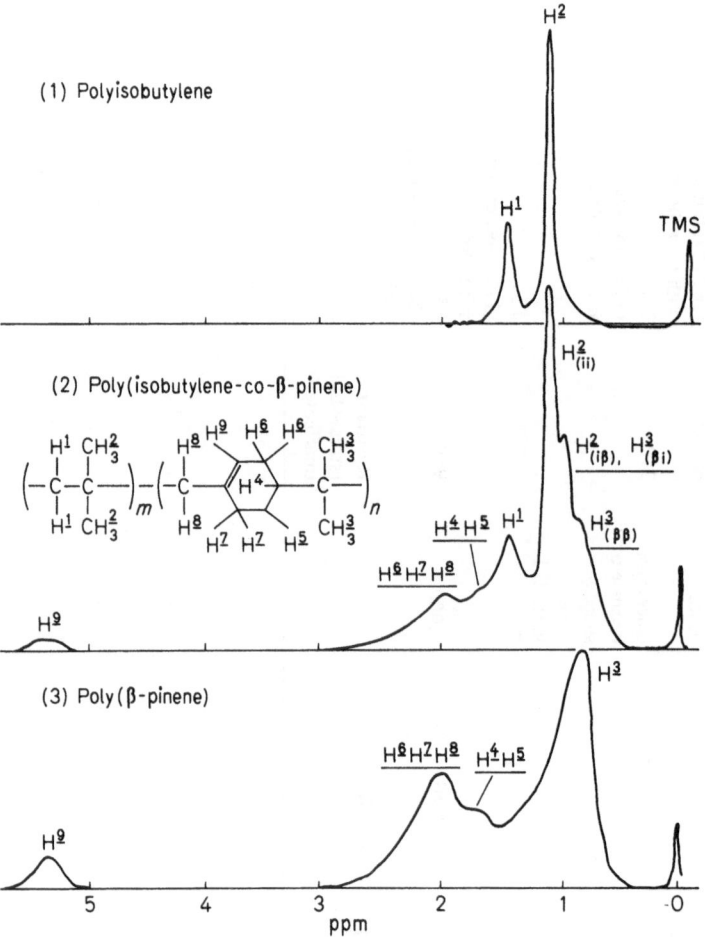

Fig. 2. PMR assignment of poly(isobutylene-*co*-β-pinene)

that does not appear in the spectra of the respective homopolymers and is therefore viewed to be due to gem-dimethyl groups in our copolymers. Consequently,
we assigned this resonance to "half-crowded" gem-dimethyl groups, *i.e.*, a gem-dimethyl group flanked by only one other gem-dimethyl group, associated with iβ or
βi sequences.

These assignments are shown schematically by the following formula:

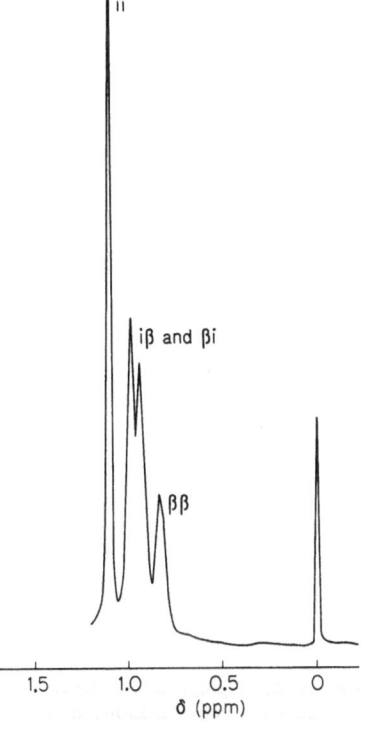

sequence:	ii	iβ		βi	iβ		ββ
chemical shift, δ:	1.1	1.0		1.0	1.0		0.8

Select copolymer samples have also been examined by high resolution (300 MHz) spectroscopy. Figure 3 shows a representative trace in the methyl region. The resonance associated with iβ or βi sequences now appears as a complicated doublet possibly suggesting pantade effects. On closer examination even the band assigned to ββ sequences is a composite of unresolved resonances. The olefinic proton resonance also reveals a composite shape. The band at 5.3δ shown in Fig. 2 is assigned to the olefinic protons in the cyclohexene structure. High resolution (300 MHz) PMR spectroscopy revealed that this resonance can be resolved into a doublet and that the shape of the doublet varies with the overall composition of the copolymer. We have not further explored this finding.

As discussed in Experimental, the overall composition of our copolymers can be calculated from the ratio of the area of the olefinic proton to the rest of the area in the spectrum.

Fig. 3. PMR spectrum of poly(isobutylene-co-β-pinene) (300 MHz; composition i = 59%, β = 41%)

The range from 1.4 to 3.0δ in the spectrum is assigned to various methylene proton resonances. The individual assignments shown in Fig. 2 become obvious upon an examination of the spectra of the respective homopolymers. Evidently the positions of the various methylene bands in the copolymer spectra are determined by their respective distances from the unsaturation. Thus protons α to the double bond appear at around 2.0δ, β protons resonate at around 1.7δ, and protons farther away appear at 1.4δ.

3. Sequence Distribution Analysis

These PMR findings prompted us to undertake a detailed sequence distribution analysis of poly(isobutylene-*co*-β-pinene). Sequence distributions can be expressed by the run number R, *i.e.*, the average number of hetero-linkages per 100 consecutive monomer linkages in a copolymer chain (*21*).

In case of poly(isobutylene-*co*-β-pinene) the characteristic resonances associated with the gem-dimethyl structures in the δ = 0.8 to 1.1 range provide an ideal diagnostic tool for the analysis of dyad distributions. The relative proportion of resonances determined at δ = 0.8, 1.0 and 1.1 reflect the relative proportion of uncrowded, half-crowded and fully-crowded gem-dimethyl groups respectively. Thus

$$A_1 : A_2 : A_3 = P_{\beta\beta} : (P_{\beta i} + P_{i\beta}) : P_{ii} \tag{4}$$

where A_1, A_2, and A_3 are the experimentally determinable relative intensities of PMR resonances associated with uncrowded, half-crowded and fully-crowded gem-dimethyl groups and $P_{\beta\beta}, P_{\beta i}, P_{i\beta}, P_{ii}$ are the probabilities or percent amounts of possible dyads in the copolymer. The subscripts β and i indicate β-pinene and isobutylene, respectively.

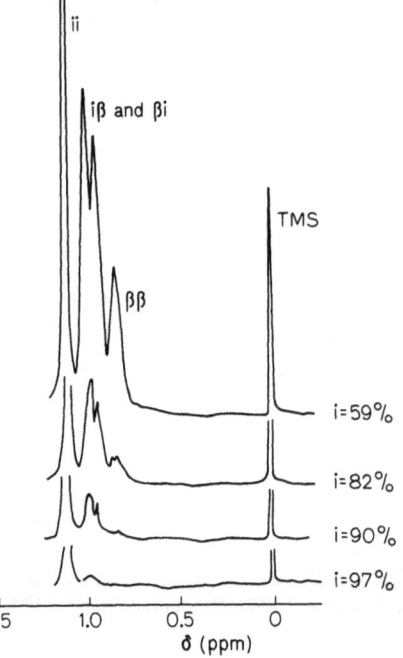

Fig. 4. PMR spectra of poly(isobutylene-*co*-β-pinenes). (Isobutylene contents shown in wt. %)

Table 2. Run numbers of poly(isobutylene-co-β-pinene)

Sample	Isobutylene content of copolymer (wt.%)	$A_1:A_2:A_3$	R
Homopolyisobutylene	100	100:0:0	0
Poly(isobutylene-co-β-pinene):			
3^a	97	94:5:1	5
4^a	90	82:13:5	15
5^a	82	58:35:8	35
6^a	59	36:48:16	48

[a] See Table 1.

The run number R is related to this ratio by Eq. (5) (21):

$$P_{\beta\beta} : (P_{\beta i} + P_{i\beta}) : P_{ii} = (100 - i\% - R/2) : R : (i\% - R/2) \tag{5}$$

where i% = concentration of isobutylene units in the copolymer. Thus, R can be obtained by determining overall copolymer composition and the respective areas at $\delta = 0.8$, 1.0, and 1.1 in its PMR spectrum.

We have used a trial and error least square method to select the best experimental value of R. The PMR spectra of a series of poly(isobutylene-co-β-pinene), containing from 59 to 97 weight% isobutylene, are shown in Fig. 4. The run numbers obtained from these spectra are listed in Table 2.

Alternatively, run numbers of copolymers can also be calculated from reactivity ratios using Eq. (6) (21):

$$R = \frac{200}{2 + r_1 \cdot \dfrac{f_1}{f_2} + r_2 \cdot \dfrac{f_2}{f_1}} \tag{6}$$

where f_1 and f_2 are the monomer mole fractions in the charge. This relationship is valid only for random copolymerizations that follow the simple "two parameter scheme" in which only four propagation steps are involved:

$$M_1^\oplus + m_1 \xrightarrow{k_{P11}} M_1^\oplus \tag{7}$$
$$M_1^\oplus + m_2 \xrightarrow{k_{P12}} M_2^\oplus \tag{8}$$
$$M_2^\oplus + m_1 \xrightarrow{k_{P21}} M_1^\oplus \tag{9}$$
$$M_2^\oplus + m_2 \xrightarrow{k_{P22}} M_2^\oplus \tag{10}$$

where $k_{P11}, k_{P12}, k_{P21}$, and k_{P22} are the rate constants of reactions (7)–(10), respectively. Thus an agreement between calculated and experimental run numbers (obtained from PMR measurements) will indicate the validity of the two parameter model to the isobutylene-β-pinene system and, consequently, the random nature of the copolymer.

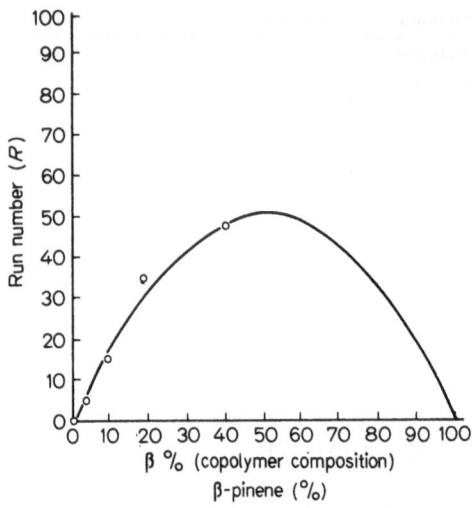

Fig. 5. Run number of poly(iso-
butylene-*co*-β-pinene) vs. copolymer composition

Fig. 5 illustrates our results. The circles are experimental data obtained from
PMR measurements whereas the cone shaped curve has been calculated by Eq. (6)
for a perfectly random isobutylene/β-pinene copolymer. Composition data for the
theoretical curve has been calculated from reactivity ratios obtained in this research
(see later). The close agreement between experimental and theoretical values is strong
indication for a random copolymer. Importantly, this close fit also indicates the
validity of the two parameter scheme for the cationic copolymerization of isobutylene and β-pinene.

4. Determination of Reactivity Ratios from Run Numbers

According to Harwood and Ritchey (*21*), the reactivity ratios of a copolymer can
be calculated by plotting $\dfrac{2(100/R - 1)(1-f_1)}{f_1}$ vs. $(1-f_1)^2 f_1^2$ where f_1 is the copolymer composition and R the run number at that composition. To gain further
insight into the detailed composition of poly(isobutylene-*co*-β-pinene) produced by
EtAlCl$_2$ in EtCl diluent at $-130\ ^\circ$C we have determined the reactivity ratios for the
isobutylene/β-pinene pair by the use of Harwood-Ritchey and conventional Fineman-Ross plots. Fig. 6 shows the Harwood-Ritchey plot and Table 3 shows a summary

Table 3. Reactivity ratios of isobutylene (r_i) and β-pinene (r_β) (Copolymerization with EtAlCl$_2$ in EtCl at -130°)

	r_1	r_β	$r_i r_\beta$
Fineman-Ross plot	1.0 ± 0.1	1.0 ± 0.1	1.0
Harwood-Ritchey plot	1.1 ± 0.1	0.9 ± 0.1	0.99

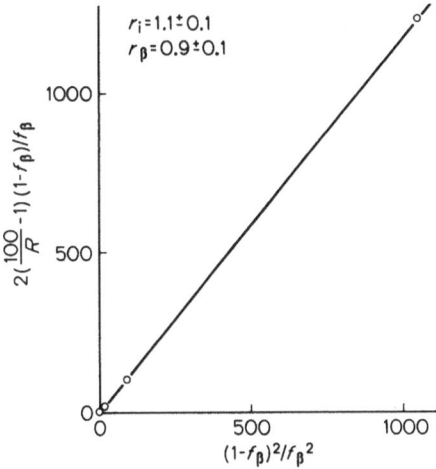

$r_i = 1.1 \pm 0.1$
$r_\beta = 0.9 \pm 0.1$

y-axis: $2(\frac{100}{R} - 1)(1 - f_\beta)/f_\beta$

x-axis: $(1 - f_\beta)^2/f_\beta^2$

Fig. 6. Harwood-Ritchey plot for isobutylene and β-pinene copolymerization (Et$_2$AlCl at $-130°$)

of the data. According to these results, cationically produced poly(isobutylene-co-β-pinenes) synthesized under the above conditions are ideal random copolymers ($r_i r_\beta \sim 1.0$) and the simple two parameter model is valid for the system (detailed reactivity ratio studies are described in the next section).

B. Reactivity Ratio Studies

A detailed study of the reactivity ratios of the isobutylene and β-pinene system has been undertaken. We found that the reactivity ratio product is close to unity over the whole temperature range studied from $-50°$ to $-130°$, indicating a random copolymer system. Further, we discovered that while the reactivity ratios are quite insensitive to the particular Lewis acid used, they can be controlled by temperature and that the individual reactivity ratios become equal to unity below about $-110°$. In other words, at very low temperatures the copolymerization becomes azeotropic (the composition of the feed and that of the copolymer become equal).

1. The Effect of Temperature on the Reactivity Ratios

The monomer charge vs. copolymer composition data for a series of isobutylene and β-pinene copolymerizations using EtAlCl$_2$ in EtCl solvent between $-50°$ and $-130°$ are shown in Fig. 7. Evidently from $-50°$ to $-100°$, β-pinene is more reactive than isobutylene, and the copolymer is relatively richer in β-pinene than the monomer charge. However, in the range from $-110°$ to $-130°$ the two monomers exhibit equal reactivities and the copolymer composition is equal to that of the monomer charge. Table 4 shows the reactivity ratios obtained together with the calculation methods employed.

Subsequently the data were treated by the Arrhenius concept, i.e., the logarithm of the reactivity ratios was plotted versus the reciprocal temperature as shown in Fig. 8. This plot expresses quantitatively the effect of temperature on the reactivity ratios of isobutylene and β-pinene. Evidently, in the higher temperature range, from $-50°$

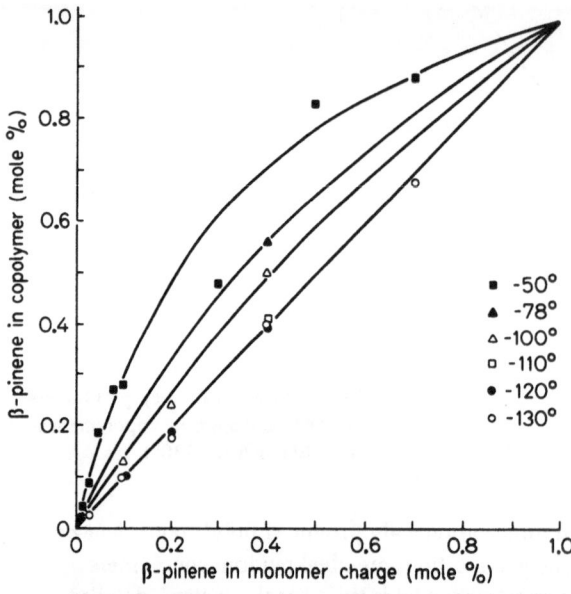

Fig. 7. Charge vs. composition plot for isobutylene-β-pinene copolymerization at various temperatures (EtAlCl$_2$ in EtCl)

to $-100\,^{\circ}$C, β-pinene is a considerably more reactive monomer than isobutylene. With decreasing temperature, the relative reactivity of β-pinene decreases while that of isobutylene increases until at $-110\,^{\circ}$C the two reactivities become equal, *i.e.*, azeotropic copolymerization conditions are reached.

A similar effect of temperature on the reactivity ratios has already been observed in cationic copolymerizations. Thus M. Imoto and K. Saotome (*23*) found in a brief

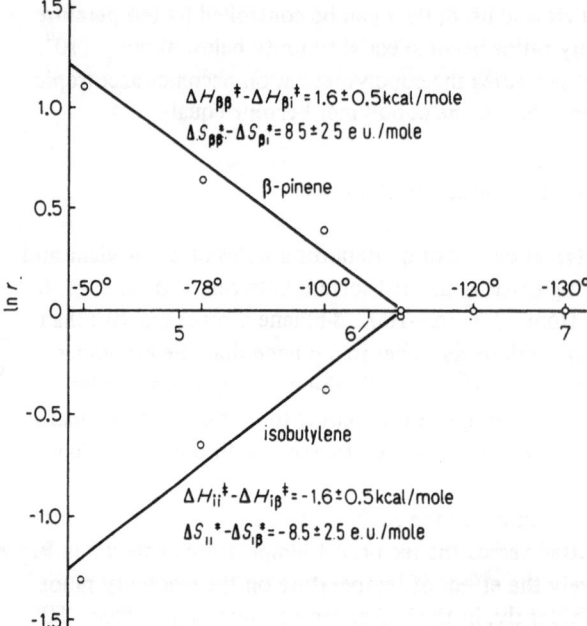

Fig. 8. The effect of temperature on the reactivity ratios of isobutylene and β-pinene (EtAlCl$_2$ in EtCl)

Table 4. The effect of temperature on the reactivity ratios of isobutylene and β-pinene

Temperature °C	r_i	r_β	$r_i \cdot r_\beta$	Calculation method
EtAlCl$_2$ in EtCl				
− 50	0.27 ± 0.06	3.0 ± 1.5	0.81 ± 1.5	a
− 78	0.52	1.9	0.99	b
−100	0.77 ± 0.16	1.5 ± 0.5	1.15 ± 0.7	c
−110	1.0	0.95	0.95	b
−120	1.0 ± 0.1	1.0 ± 0.1	1.0 ± 0.1	c
−130	1.0 ± 0.1	1.0 ± 0.1	1.0 ± 0.1	a
AlCl$_3$ in MeCl				
− 60	0.4 ± 0.1	2.3 ± 0.5	0.9 ± 0.6	c
−100	1.1 ± 0.1	1.05 ± 0.1	1.15 ± 0.2	c
−100	1.0 ± 0.1	0.93 ± 0.2	0.93 ± 0.3	a
BF$_3$ in MeCl				
− 70	0.92	1.1	0.97	b
−100	5.4	0.2	1.08	b

[a] By Fineman-Ross method and confirmed by the Mortimer-Tidwell (22) non-linear least square computed method.
[b] By assuming that the product of reactivity ratio equals unity and using the relationship: $r_i = F_i \cdot f_\beta / F_\beta \cdot f_i$ where F_i and f_i denote the mole fractions of isobutylene in the copolymer and charge, respectively, and F_β and f_β those for β-pinene.
[c] By Fineman-Ross method.

study that in the copolymerization of acenaphthylene (AcN) and n-butyl vinyl ether (nBuVE) (with $BF_3 \cdot OEt_2$ in benzene or toluene in the range from $-30°$ to $-78\,°C$) the product of reactivity ratios was approximately unity and that the r_{nBuVE} increased whereas that of r_{AcN} decreased with decreasing temperatures. The plot of $\log r_{AcN}$ vs. $1/T$ gave a straight line, the slope and intercept of which gave the differences of the enthalpy and entropy of activation.

It has been suggested that the linear $\log r$ vs. $1/T$ plot passes near to the origin in free radical copolymerization (24). However, it appears that $\log r$ vs. $1/T$ plots do not pass through the origin in cationic copolymerization [(25), p. 571 of Ref. (11)]. This means that the activation entropy for the addition of different monomers to the same propagating end is the same for free radical copolymerization, but different for cationic copolymerization. The result of our work is in line with the later conclusion. Furthermore, the effect of temperature on reactivity ratio found in our study i.e., reactivity ratio approaches unity at low temperatures is just the opposite to that of free radical copolymerization (24) where the reactivity ratio approaches unity at high temperatures. Evidently, free radicals become less and less selective that is more aggressive with increasing temperature while the cations of isobutylene and β-pinene in the moderately polar solvent become increasingly aggressive with decreasing temperature. Possibly this is due to the dielectric constant increasing effect of decreasing temperatures: Decreasing temperatures promote solvation and consequently increase the concentration of free, counterion unencumbered, more aggressive carbocations.

Quantitatively, the differences in activation entropies and enthalpies between homo- and cross-propagations can be calculated from the intercepts and slopes in Fig. 8, respectively:

$$S^{\neq}_{p_{ii}} - S^{\neq}_{p_{i\beta}} = -8.5 \pm 2.5 \text{ e.u./mole}$$

$$S^{\neq}_{p_{\beta\beta}} - S^{\neq}_{p_{\beta i}} = +8.5 \pm 2.5 \text{ e.u./mole}$$

$$H^{\neq}_{p_{ii}} - H^{\neq}_{p_{i\beta}} = -1.6 \pm 0.5 \text{ kcal/mole}$$

$$H^{\neq}_{p_{\beta\beta}} - H^{\neq}_{p_{\beta i}} = +1.6 \pm 0.5 \text{ kcal/mole}$$

where $S^{\neq}_p$ and $H^{\neq}_p$ are activation entropy and enthalpy of propagation, respectively.

These date provide some insight into the details of the propagation and isomerization step. Thus isomerization might proceed by a two step mechanism separated by a transition state:

or by a "concerted" mechanism via an activated complex:

According to Dainton and Ivin (26) the polymerization entropy of small cyclic monomers (including four membered rings) in generally higher (more favorable) than
that of ethylenic monomers. This fact combined with our finding that in terms of
the activation entropy, the addition of β-pinene is more favorable than that of isobutylene might indicate that β-pinene isomerization copolymerization and, by ana-
logy, the isomerization homopolymerization of β-pinene proceed by a "concerted"
mechanism involving a partially opened ring system with some extra degrees of
freedom. Kinetically, the two step mechanism is equivalent to a conventional vinyl
polymerization because intramolecular rearrangement (step 2) must be faster than
intermolecular monomer addition (step 1) which therefore becomes rate-determin-
ing.

2. The Effect of Counter-Ion on the Reactivity Ratios

While most of our investigations have been carried out with $EtAlCl_2$, a limited
amount of work was also done with $AlCl_3$ and BF_3. Experimental conditions, calculation methods and reactivity ratios are shown in Table 4. The trends are similar
to those observed with $EtAlCl_2$: a) r_i increases while r_β decreases with decreasing
temperatures b) r_i and r_β are close to unity at $-100°$ c) the product $r_i \cdot r_\beta$ are close
to unity at every temperature level.

3. The Effect of Monomer Concentration on the Reactivity Ratios

In the course of this research we have also briefly investigated the effect of monomer
concentration on the reactivity ratios of isobutylene and β-pinene. Three experiments
have been carried out using 60/40 (mole percent) isobutylene/β-pinene charges with
$EtAlCl_2$ in EtCl at $-110°$ and 8.3, 5.0 and 2.8 mole/l total monomer concentrations.
Concentrations did not seem to affect reactivity ratios under the conditions employed.

C. The Effect of Temperature and Charge Composition on Copolymer Molecular Weight and Initiator Efficiency

1. Isobutylene-β-Pinene System

The effect of temperature and relative monomer charge composition on the molecular weights of poly(isobutylene-co-β-pinene) has been investigated. The molecular
weight increasing effect of decreasing temperature is well known in cationic polymerizations (27). Although very high molecular weight polyisobutylenes can readily
be obtained at, say, $-100\,°C$, the molecular weights of the isobutylene-β-pinene
copolymers dropped precipitously even in the presence of relatively small amounts
(10–20%) of β-pinene. Preliminary studies have shown that unless the copolymerization is carried out at very low temperatures (e.g. $< -100\,°C$), low molecular
weight, semiliquid, viscous, tacky copolymers are obtained. High molecular weight

copolymers were synthesized at very low temperatures *e.g.*, −130 °C (see Table 1). Fig. 9 shows the effect of the charge composition on the viscosity average degrees of polymerization of poly(isobutylene-*co*-β-pinene) obtained at −130 °C with EtAlCl$_2$.

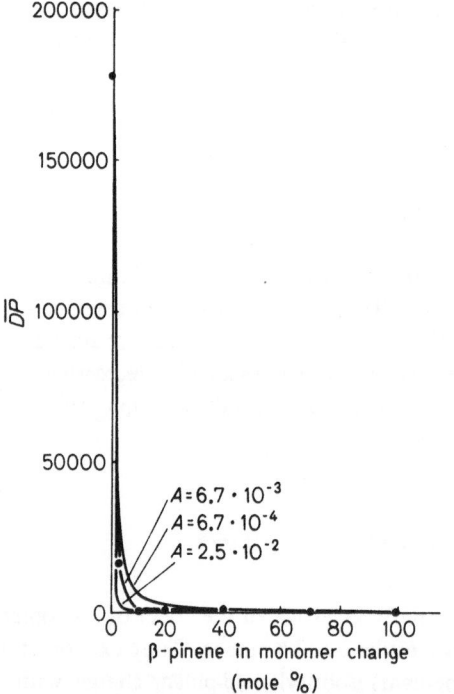

Fig. 9. The effect of monomer charge on the molecular weight of poly-(isobutylene-*co*-β-pinene), EtAlCl$_2$ in EtCl at −130° (Higashimura's *A* value indicated, see text for explanation).

The molecular weight depressing effect of even small amounts of β-pinene is remarkable: The $\overline{DP}_v$ of a copolymer containing ∼ 3 and 10 mole% β-pinene are, respectively, about one and two orders of ten lower than the $\overline{DP}_v$ of polyisobutylene.

The question as to be cause of this effect immediately arises. A thorough literature search revealed that while there were numerous authors who observed the phenomenon (see Table 5) of molecular weight depression in cationic copolymerizations, there was only one group of workers who treated this problem in some depth. In a series of papers Higashimura *et al.* (28−31) examined the molecular weights of poly(isobutylene-*co*-α-methylstyrene) and found, similarly to our observations with β-pinene, that the molecular weight of the copolymer drops rapidly in the presence of even modest amounts of styrene or α-methylstyrene and that the molecular weights of the copolymers containing appreciable amounts of either components were always lower than those of the respective homopolymers.

Higashimura developed a kinetic theory to explain these observations. We have adopted and further developed Higashimura's work to explain our data and to derive some important generalizations.

The molecular weight of any cationically synthesized polymer is governed by the ratio:

$$DP = \frac{\text{rate of propagation}}{\text{rate of (transfer + termination)}}.$$

Hence the degree of polymerization, DP of a copolymer is given by

$$DP = \frac{k_{p_{11}}[M_1^\oplus][m_1] + k_{p_{12}}[M_1^\oplus][m_2] + k_{p_{22}}[M_2^\oplus][m_2] + k_{p_{21}}[M_2^\oplus][m_1]}{k_{t_1}[M_1^\oplus] + k_{tr_{11}}[M_1^\oplus][m_1] + k_{tr_{12}}[M_1^\oplus][m_2] + k_{tr_{22}}[M_2^\oplus][m_2] + k_{tr_{21}}[M_2^\oplus][m_1] + k_{t_2}[M_2^\oplus]}$$

(11)

where the k_p, k_{tr}, k_t are the rate constants of propagation, transfer to monomer and unimolecular termination respectively; $M^\oplus$ and m signify the propagating carbenium ion and monomer respectively. The subscripts 1 and 2 refer to monomers used. It is assumed that the only type of transfer that operates is transfer to the monomer.

Since carbenium ions are extremely reactive intermediates and the molecular weights of cationically produced polymers are by and large determined by chain transfer to monomer, one can neglect k_{t_1} and k_{t_2} in the denominator. Thus by rearranging Eq. (11), one obtains:

$$DP_{12} = \frac{r_1[m_1]^2 + 2[m_1][m_2] + r_2[m_2]^2}{\frac{1}{DP_1} \cdot r_1 \cdot [m_1]^2 + A[m_1][m_2] + \frac{1}{DP_2} \cdot r_2 \cdot [m_1]^2}$$

(12)

where DP_{12}, DP_1 and DP_2 are, respectively, the degree of polymerization of copolymers and that of the homopolymers from monomer 1 and 2, and $A = k_{tr\,12}/k_{p\,12} + k_{tr\,21}/k_{p\,21}$. "$A$" determines the extent of molecular weight decrease of the copolymer relative to the homopolymers.

By a trial and error curve fitting method Higashimura obtained A values for the isobutylene-styrene and isobutylene-α-methylstyrene systems. For example, $A = 6.9 \times 10^{-3}$, with DP(polyisobutylene) = 2,000 and DP(polystyrene) = 5,500 (TiCl$_4$ initiator, hexane/CH$_2$Cl$_2$:l/l, at $-78\,^\circ$C) (28). In this case, $A \gg 1/DP_1$ or $1/DP_2$, i.e., A is larger than either of the two reciprocals because the degrees of polymerization of both homopolymers are similar. In cases where $DP_1 \gg DP_2$, A will approach $1/DP_2$ but it can never approach $1/DP_1$. The Japanese authors recognized that one term in A must be predominant while the other is insignificant and can be neglected.

We employed Higashimura's method to calculate an A value for the isobutylene-β-pinene system. Data are shown in Fig. 9. The necessary DP values have been estimated from available $\overline{M}_v$ data by dividing by the average monomer molecular weight [β-pinene content] x 136 + [isobutylene content] x 56 g/mole. The errors in $\overline{M}_v$ are expectedly large, since the equation we used to obtain $\overline{M}_v$ is strictly valid only for polyisobutylene (19). However, the error introduced by this approximation should be insignificant in view of the order of magnitude of molecular weight drop. A is thus estimated to be 6.7×10^{-3} as shown in Fig. 9. Since in this case $DP_1 \gg DP_2$, the value of A approaches $1/DP_2$

2. A General Treatment of Molecular Weight Depression in Cationic Copolymerization

We have expanded and built upon Higashimura's treatment by examining the general significance of molecular weight reduction in cationic copolymerization. We will demonstrate that *by introducing a second copolymerizable monomer into a cationic homopolymerization, the molecular weight of the copolymer must decrease in reference to the respective molecular weights of the homopolymers provided a) both homopolymers produce high polymers and b) the copolymer contains appreciable amounts (e.g. 5%) of each monomer.* If monomer m_2 is able to copolymerize with monomer m_1, one of the cross transfer reactions must be much faster than either of the homo-transfer reactions, thus it will dominate the denominator of Eq. (11) and consequently will depress the molecular weight of the copolymer. The following analysis should illuminate these statements.

In a copolymerization, there must coexist two propagating chain ends of differing energies. Given this situation and keeping in mind Hammond's postulate for exothermic reactions (*32*), the fastest event will be the one which involves the most active chain end, say $M_1^\oplus$ and the most stable monomer, say m_2, and will produce either the most stable chain end $\sim\sim M_2^\oplus$ (cross-propagation) or most stable cation $m_2^\oplus$ plus terminated macroolefin $\sim\sim M_1^= $ = (cross-transfer),

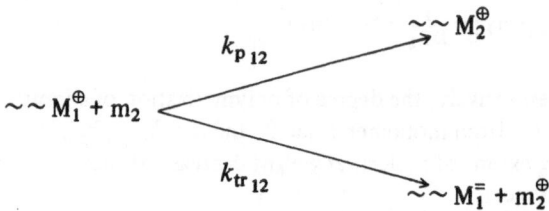

i.e., the fastest propagation reaction will be $k_{p\,12}$ and the fastest transfer reaction $k_{tr_{12}}$. The molecular weight of the copolymer will be determined by the competition between $k_{p_{12}}$ and $k_{tr\,12}$ *i.e.* cross-propagation and cross-transfer.

We recall in cationic homopolymerizations $E_{tr_{11}}^{\neq} \gg E_{p_{11}}^{\neq}$, *i.e.*, the activation energy for homotransfer $E_{tr_{11}}$ is much larger than that of homopropagation $E_{p\,11}$ and that the latter might indeed be close to zero as it involves an ion-molecule reaction. It is not too far fetched to assume that this is also true for copolymerizations. Therefore, by lowering the energy content of the monomer from m_1 in a homopolymerization to m_2 in a copolymerization, the activation energy of cross-transfer $E_{p_{12}}^{\neq}$ decreases *more* than that of cross-propagation $E_{tr_{12}}^{\neq}$ (*i.e.*, $-\Delta E_{tr}^{\neq} > -\Delta E_p^{\neq}$) because the value of the latter is very small (0–1 kcal/mole) to begin with. These relationships are illustrated in Fig. 10. According to Hammond's postulate, then, *the rate of cross-transfer must have increased more than that of cross-propagation which must result in decreased molecular weights.*

A) Schemactic showing of the lowering of activation
   energy in the favored cross event

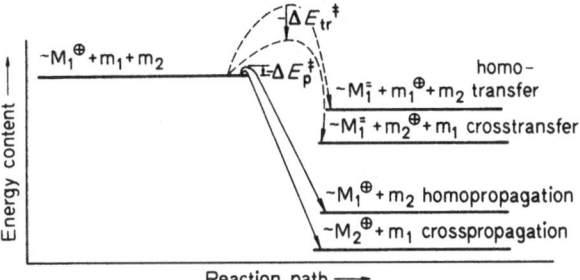

B) Schemactic showing of the increasing of activation
   energy in the not favored cross event

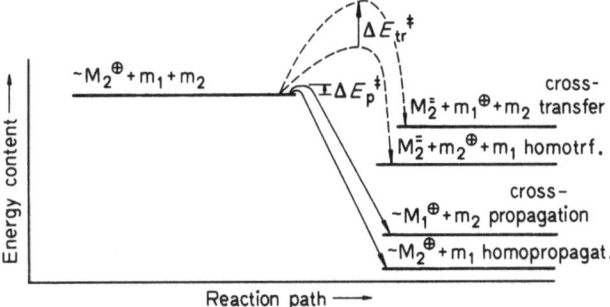

Fig. 10. Idealized energy diagram for a cationic copolymerization

On the other hand, by increasing the energy content of the monomer from m_2 in a homopolymerization to m_1 in a copolymerization, the activation energy of cross-transfer $E^{\neq}_{tr_{21}}$ increases *more* than that of cross-propagation $E^{\neq}_{p_{21}}$ i.e., $\Delta E^{\neq}_{tr} >$ $> \Delta E^{\neq}_{p}$ which would result in increased molecular weight. Obviously this latter scenario cannot be true simply because $k_{tr_{12}}$ dominates the denominator in Eq. (11) and $k_{tr_{21}}$ is negligible.

Based on this analysis we have reached the following conclusions:

a) The molecular weight of a random copolymer produced by carbenium ion mechanism in which $E^{\neq}_{tr} > E^{\neq}_{p}$, will be lower than that of the respective homopolymers obtained under identical conditions

b) The phenomenon of molecular weight depression in copolymerization is pronounced at low temperatures because the term $e^{-\Delta E^{\neq}/RT}$ becomes larger with decreasing temperatures.

c) This molecular weight decreasing phenomenon will be small if both copolymerizable monomers produce low molecular weight homopolymers because in such systems the activation energy difference between propagation and transfer is small. For example, in the copolymerization system of styrene and p-chlorostyrene with $HClO_4$ initiator at $25°$ (*33*), a molecular weight depressing was not observed because of the very low molecular weights ($\overline{Mn}$ = 1360 to 1460) obtained over the whole monomer composition range (0 to 0.77 mole fraction of styrene). Under these conditions $E_{tr}^{\neq}$ was not higher than $E_{p}^{\neq}$. For the same reason, this effect will be large if the respective monomers produce high molecular weight homopolymers.

   d) On the basis of a similar analysis in regard with the importance of unimolecular termination as the polymer yield determining event, we conclude that initiator efficiency (grams of polymer produced per mole of initiator) should also decrease precipitously with increasing amount of comonomer in the charge. Our data shown in Fig. 11 substantiate this conclusion.

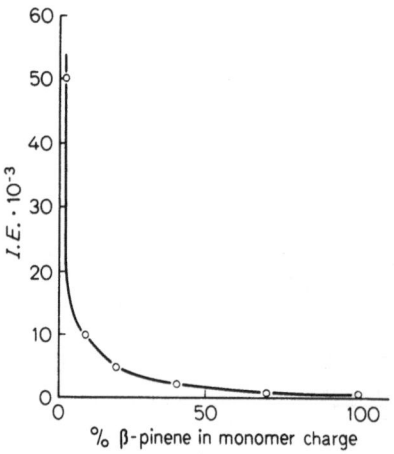

Fig. 11. The effect of monomer feed composition on initiator efficiency of poly (isobutylene-*co*-β-pinene)

   A thorough literature search to test our conclusions has been undertaken. Systems that are in agreement with our theory are shown in Table 5 and we have failed to uncover a copolymerization system which would not follow our analysis.

V. Summary and Conclusions

Mechanistic-structural considerations prompted us to investigate the cationic copolymerization of isobutylene and β-pinene. In contrast to claims in the literature (*17*), these two monomers can be readily copolymerized to relatively high molecular weight products by the use of EtAlCl$_2$ in ethyl chloride diluent in the −50° to −130 °C range. According to gel permeation chromatography the copolymers are homogeneous, *i.e.*, show monomodal molecular weight distribution (Fig. 1). This is contrary to Finnish author's claims (*17*) that only mixtures of homopolymers are produced. The discrepancy between these findings has been analyzed and resolved.

   We have analyzed the PMR spectra of our copolymers. In the methyl proton region the three partially resolved resonances at 1.1, 1.0 and 0.8δ have been assigned, respectively, to gem-dimethyl groups flanked by two gem-dimethyl groups ("fully crowded gem-dimethyls"), gem-dimethyl groups flanked by only one other dem-dimethyl group ("half-crowded gem-dimethyls"), and gem-dimethyl groups not flanked

Table 5. Cationic copolymerization systems in which molecular weight depression has been observed

Monomers	Initiator	Solvent	Temp., °C	Ref.
Isobutylene/β-pinene[a]	EtAlCl$_2$	EtCl	−50° to −130°	This work
Isobutylene/styrene	TiCl$_4$	CH$_2$Cl$_2$, toluene	− 78°	(29, 31)
Isobutylene/α-methylstyrene	TiCl$_4$	CH$_2$Cl$_2$, toluene	− 78°	(29, 31)
Isobutylene/isoprene	BF$_3$	Ethylene, EtCl	−100°	(34)
Isobutylene/butadiene	BF$_3$	Ethylene, EtCl	−100°	(34)
Isobutylene/2,3,dimethylbutadiene	BF$_3$	Ethylene, EtCl	−100°	(34)
Isobutylene/2-methyl-2-butene	TiCl$_4$	CH$_2$Cl$_2$, hexane	− 78°	(34)
Isobutylene/cyclopentadiene	BF$_3$ · O(Et)$_2$	CH$_2$Cl$_2$, toluene	− 78°	(36)
Isobutylene/cyclohexadiene	BF$_3$ · O(Et)$_2$	CH$_2$Cl$_2$, toluene	− 78°	(36)
Isobutylene/acenaphthylene	AlCl$_3$, BF$_3$ · O(Et)$_2$	EtBr, CS$_2$, CHCl$_3$	−60° to −105°	(37)
Styrene/α-methylstyrene	TiCl$_4$ · CCl$_3$CO$_2$H	CH$_2$Cl$_2$, toluene	− 78°	(29, 31)
Styrene/acenaphthylene[a]	BF$_3$ · O(Et)$_2$	Benzene, toluene	−30° to −78°	(38)
Styrene/acenaphthylene	TiCl$_4$	CH$_2$Cl$_2$	− 72°	(39)
Styrene/1-methylacenaphthylene	TiCl$_4$	CH$_2$Cl$_2$	− 72°	(39)
Styrene/3-methylacenaphthylene	TiCl$_4$	CH$_2$Cl$_2$	− 72°	(39)
Styrene/5-methylacenaphthylene	TiCl$_4$	CH$_2$Cl$_2$	− 72°	(39)
Styrene/2-methyl-2-butene	TiCl$_4$	CH$_2$Cl$_2$, hexane	− 78°	(35)
Styrene/chloroprene	BF$_3$ · O(Et)$_2$	Cyclohexane	− 18°	(40)
Styrene/5-methylindene	TiCl$_4$	CH$_2$Cl$_2$	− 50°	(41)
Indene/cumarone	TiCl$_4$	CH$_2$Cl$_2$	− 72°	(42)
Indene/5,7-dimethylindene	TiCl$_4$	CH$_2$Cl$_2$	− 30°	(43)
Indene/4,6,7-trimethylindene	TiCl$_4$	CH$_2$Cl$_2$	− 72°	(43)
Indene/7-methylindene	TiCl$_4$	CH$_2$Cl$_2$	− 70°	(44)
Indene/4,6-dimethylindene	TiCl$_4$	CH$_2$Cl$_2$	− 72°	(45)
Indene/5,6-dimethylindene	TiCl$_4$	CH$_2$Cl$_2$	− 72°	(45)
Indene/4,7-dimethylindene[a]	TiCl$_4$, BF$_3$ · O(Et)$_2$	CH$_2$Cl$_2$	− 72°	(45)
Indene/4,5,6,7-tetramethylindene[a]	TiCl$_4$, H$_2$SO$_4$	CH$_2$Cl$_2$	− 72°	(45)
Indene/1-methylindene	TiCl$_4$	CH$_2$Cl$_2$	− 72°	(46)
Indene/2-methylindene	TiCl$_4$	CH$_2$Cl$_2$	− 72°	(46)

Table 5 (continued)

Monomers	Initiator	Solvent	Temp., °C	Ref.
Indene/3-methylindene	$TiCl_4$	CH_2Cl_2	$-72°$	(46)
Indene/various methylindenes	$TiCl_4$, $BF_3 \cdot O(Et)_2$ H_2SO_4	CH_2Cl_2	$-72°$	(47)
Indene/4,7-dimethylbenzofuran[a]	$TiCl_4$	CH_2Cl_2	$-78°$	(48)
4,6-dimethylindene/4,7-dimethylindene	$TiCl_4$	CH_2Cl_2	$-72°$	(45)
4,7-dimethylindene/4,5,6,7-tetramethylindene	$TiCl_4$	CH_2Cl_2	$-72°$	(45)
Cyclopentadiene/α-methylstyrene	$BF_3 \cdot O(Et)_2$	Toluene	$-78°$	(49)
Cyclopentadiene/ 2-Chloroethyl vinyl ether	$BF_3 \cdot O(Et)_2$	CH_2Cl_2, $EtNO_2$, Toluene	$-78°$	(50)
α-Methylstyrene/2-Chloroethyl vinyl ether	$BF_3 \cdot O(Et)_2$	CH_2Cl_2	$-78°$	(51)
p-Methoxystyrene/2-Chloroethyl vinyl ether	$BF_3 \cdot O(Et)_2$	CH_2Cl_2	$-78°$	(51)
Benzofuran/4,7-dimethylbenzofuran[a]	$TiCl_4$	CH_2Cl_2	$-78°$	(48)

[a] Depression in yield was also observed.

by neighboring gem-dimethyl group ("uncrowded gem-dimethyls"). These assignments are proposed to be diagnostic for isobutylene copolymers in general. The resonance at 5.3 δ was assigned to olefinic protons in poly(isobutylene-co-β-pinene) and was used to determine overall copolymer composition.

The sequence distribution of isobutylene and β-pinene units in the copolymers was elucidated by an analysis of the gem-dimethyl chemical shifts. "Run numbers" have been calculated and found to coincide with theoretical values for statistical copolymers (Fig. 5).

Reactivity ratios of isobutylene and β-pinene have been determined. The product of reactivity ratios is approximately unity and, interestingly, r_β decreases while r_i increases with decreasing temperatures in the range from -50 to $-110°$. The plot of the logarithm of reactivity ratios versus the reciprocal temperature gave a straight line, the slope and intercept of which yielded the difference of enthalpy and entropy of activation (Fig. 8). Thus, β-pinene is more reactive in terms of activation entropy and isobutylene is more reactive in terms of activation enthalpy. Significantly, below $\sim -100°$ both reactivity ratios become equal to unity, i.e., azeotropic copolymerization prevails.

Copolymer molecular weights were determined by viscometry and osmometry. Depending on the amount of β-pinene incorporation, the molecular weights range from $\sim 10^6$ for low β-pinene content copolymers (e.g., $\beta\% = 3$ mole%) to $\sim 10^3$ for moderate β-pinene content copolymers (e.g., $\beta\% = 41$ mole%). The molecular weight of poly(isobutylene-co-β-pinene) is much lower than that of polyisobutylene even in the presence of relatively small amounts(10–20%) of β-pinene (Fig. 9). We have examined in detail the phenomenon of molecular weight depression in cationic copolymerization and concluded that copolymer molecular weight must decrease in reference to the homopolymer molecular weight because cross-transfer competes favorably with cross-propagation in cationic copolymerization. Our literature search has failed to uncover a cationic copolymerization system which would not follow our analysis.

Acknowledgement. Financial assistance by the National Science Foundation is gratefully acknowledged.

VI. References

1. Roberts, W. J., and Day, A. R.: J. Am. Chem. Soc. **72**, 1226 (1950).

2. Huet, J. M., and Marechal, E.: C. R. Acad. Sci. Paris, **271**, 1058 (1970).

3. Marvel, C. S., Hanley, J. R., and Longone, D. T.: J. Polym. Sci. **55**, 551 (1959).

4. Achon, M. A., Garia-Banon, M. I., and Matco, J. L.: Makromol. Chem. **26**, 175 (1958).

5. Bates, T. H., Best, J. V. F., and Williams, T. F.: J. Chem. Soc. **298**, 1531 (1962).

6. Van Lohuizen, O. E., and DeVries, K. S.: J. Polym. Sci. **C16**, 3943 (1968).

7. Kennedy, J. P., and Makowski, H. S.: J. Makromol. Sci. Chem. **A1**, 345 (1967).

8. Kennedy, J. P., and Makowski, H. S.: J. Polym. Sci. **C22**, 247 (1968).

9. Guterbock, H.: Polyisobutylene. Berlin – Göttingen – Heidelberg: Springer 1959.

10. Kennedy, J. P., and Kirshenbaum, I.: Vinyl and Diene Monomers E. C. Leonard, Chap. III, Part II, New York: Interscience 1971.

11. Plesch, P. H.: The chemistry of cationic polymerization. New York: Macmillan 1963.
12. Kennedy, J. P.: Cationic polymerization of olefins: A critical inventory. New York: Interscience Publishers 1971.
13. Kennedy, J. P.: Polymer chemistry of synthetic elastomers ed J. P. Kennedy and E. Tornquist, Chap. V, Part I., New York: Interscience Publishers 1968.
14. Modena, M., Bates, R. B., and Marvel, C. S.: J. Polym. Sci. A3, 949 (1965).
15. Pietrilla, H., Sivola, A., and Scheffer, H.: J. Polymer Sci. A1, 727 (1970).
16. Ott, E.: US Patent 2, 373, 706 (1945).
17. Sivola, A., and Harva, O.: Somen Kemistilehti B, 43, 476 (1970).
18. Kennedy, J. P., and Thomas, R. M.: Advances in Chemistry Series 34, 111 (1962).
19. Flory, P. J.: J. Am. Chem. Soc. 65, 372 (1943).
20. Booth, A. B.: Encycl. chem. tech. 19, 813, New York: Interscience Publishers 1969.
21. Harwood, H. J., and Ritchey, W. M.: J. Polym. Sci. B6, 277 (1964).
22. Tidwell, P. W., and Mortimer, G. A.: J. Macromol. Sci.-Revs. Macromol. Chem. C4 (2), 281 (1970).
23. Imoto, M., and Saotome, K.: J. Polym. Sci. 31, 208 (1958).
24. Alfrey, T., Jr., Bohrer, J. J., and Mark, H.: Copolymerization, high polym., Vol. VIII, p. 109. New York: Interscience 1952.
25. Masuda, T., and Higashimura, T.: Polymer J. 2, 29 (1971).
26. Dainton, F. S., and Ivin, K. J.: Quart. Rev. (London) 12, 61 (1958).
27. Thomas, R. M., Sparks, W. J., Frolich, P. K., Otto, M., and Mueller-Cunradi: J. Am. Chem. Soc. 62, 276 (1940).
28. Imanishi, Y., Higashimura, T., and Okamura, S.: J. Polym. Sci. A3, 2455 (1965).
29. Okamura, S., Higashimura, T., and Takeda, K.: Kobunshi Kagaku 18, 389 (1961).
30. Imanishi, Y., Momiyama, Z., Higashimura, T., and Okamura, S.: Kobunshi Kagaku 20, 369 (1963).
31. Okamura, S., Higashimura, T., Imanishi, Y., Yamamoto, R., and Kimura, K.: J. Polym. Sci. C16, 2365 (1967).
32. Hammond, G. S.: J. Am. Chem. Soc. 77, 334 (1955).
33. Brown, G. P., and Pepper, D. C.: Polymer 6, 497 (1965).
34. Anosov, V. I., and Korotkov, A. A.: Vysokomoek. Soed. 2, 354 (1960).
35. Imanishi, Y., Imamura, H., Higashimura, T.: Kobunshi Kagaku 27, 242 (1970).
36. Imanishi, Y., Yamane, T., Momiyama, Z., and Higashimura, T.: Kobunshi Kagaku 23, 152 (1966).
37. Jones, J. I.: J. Appl. Chem. 1, 568 (1951).
38. Saotome, K., and Imoto, M.: Kobunshi Kagaku 15, 368 (1958).
39. Belliard, P., and Maréchal, E.: Bull. Soc. Chim. France 1972, 4255.
40. Foster, F. C.: J. Polym. Sci. 5, 369 (1950).
41. Tortai, J. P., and Maréchal, E.: Bull. Soc. Chim. France 1971, 2673.
42. Sigwalt, P.: J. Polym. Sci. 52, 15 (1961).
43. Anton, A., Zwegers, J., and Maréchal, E.: Bull. Soc. Chim. France 1970, 1466.
44. Gailloud, P., Huet, J. M., and Maréchal, E.: Bull. Soc. Chim. France 1970, 1473.
45. Maréchal, E., and Evrard, P.: Bull. Soc. Chim. France 1969, 2039.
46. Maréchal, E., Basselier, J. J., and Sigwalt, P.: Bull. Soc. Chim. France 8, 1740 (1964).
47. Maréchal, E.: Kinetics and Mechanism of Polyreactions, Int. Symp. Macromol. Chem., Budapest 1, 363 (1969).
48. Zaffran, C., and Maréchal, E.: Bull. Soc. Chim. France 1970, 3523.
49. Imanishi, Y., Hara, K., Kohjiya, S., and Okamura, S.: J. Makromol. Sci. Chem. A2, 1423 (1968).
50. Kohjiya, S., Nakamura, K., and Yamashita, S.: Angew. Makromol. Chem. 27, 189 (1972).
51. Masuda, O., Higashimura, T., and Okamura, S.: Polymer J. 1, 19 (1970).

Received October 26, 1975

Poly(isobutylene - *co*-β-Pinene) A New Sulfur Vulcanizable, Ozone Resistant Elastomer by Cationic Isomerization Copolymerization

II. Characterization and Physical Properties

Joseph P. Kennedy and Tom Chou*

The Institute of Polymer Science, The University of Akron, Akron, Ohio 44325, U.S.A.

Table of Contents

* Tom Chou present address: Arlon Products, 19200 Laurel Park Road, Compton, California 90224, U.S.A.

I. Introduction

As discussed in the introduction to the first paper of this series, one of the objectives of this study was the synthesis of a series of novel high molecular weight copolymers of isobutylene and β-pinene having useful physical properties. The physical properties of these copolymers were to reflect "soft" isobutylene and "hard" β-pinene structural units.

soft   hard

As the foregoing discussion has already indicated, we have partially reached our goals. We have prepared high molecular weight isobutylene-rich poly(isobutylene-co-β-pinenes), however, we could not synthesize the high molecular weight β-pinene-rich copolymer. Indeed, the physical properties of β-pinene-rich copolymers were so poor (brittleness on account of low molecular weights) that property characterization studies have not been carried out.

In contrast, the isobutylene-rich rubbery copolymers exhibited useful properties *i.e.*, vulcanizability with sulfur and resistance to ozone, and these products have been further characterized.

II. Experimental

Copolymer syntheses have been discussed in detail in the first paper of this series. All the copolymers studied are described in Table 1 of that paper.

Glass transition temperatures were measured using a Perkin-Elmer Differential Scanning Calorimeter, Model DSC-1B. Heating rate was at $10°/min$. Accuracy was about $5°$. The instrument was calibrated by using alkanes of various melting points.

Polymers were compounded on a cold 3 x 10 inch laboratory mill using the following recipe (in parts): gum 100, stearic acid 1, ZnO 5, tetramethylthiuram disulfide 1, ALTAX 1, sulfur 1.75 and PBNA 0.5. All curatives were used as supplied without further purification. The compound became homogeneous after about forty minutes of milling. More time was required when carbon black was used. The compound was sheeted and cured in a 0.0625 inch thick mold at $160°$ for 30 min.

Vulcanization rate was determined by a Monsanto Rheometer. The Monsanto Oscillating Disk Rheometer is designed to measure the complete curing characteristics of a single rubber specimen, heated and maintained under continuous pressure during vulcanization. Sinusoidal oscillation of a conical disk, embedded in the vulcanizable mix confined in a heated square cavity, exerts a shear strain on the specimen. The force (torque) needed to oscillate the disk is directly proportional to the stiffness (shear modulus) of the specimen. As the modulus increases during vulcanization the torque is recorded against time (on a flat bed recorder) to give a curve of stiffness vs. time.

Stress-strain measurements were performed on an Instron Tester, using ASTM D412 size D dumbbells. The ultimate tensile strength of unfilled vulcanizates were difficult to determine with standard dumbbells because of excessive slipping. This difficulty was overcome by decreasing the actual load on the dumbbells, by reducing the center width to one fourth of the specifications (from 0.125 inches to 0.031 inches). Strain rate was 10 cm/min.

Shore A hardness was measured by a type A2 Indentation Durometer.

The ozone test was performed by the courtesy of Mr. M. Roxbury of this Institute. The samples were stretched to 100% extension and hung on a clamp inside the ozone chamber. The ozone concentration was measured by iodometric titration, following the procedury by C. W. Wadelin (1). The flow rate of ozone through the ozone chamber was adjusted to 25 liter per hour, the measured concentration of ozone was 1,400 ppm.

The covulcanization experiment was carried out as follows: On a laboratory mill were mixed (in parts) poly(isobutylene-co-β-pinene) (sample 4 in Table 1 of the first paper of this series) 30, butadiene rubber (Goodrich Ameripol CB220) 70, HAF carbon black 40, zinc oxide 5, N-cyclohexyl-2-benzothiazole sulfamide 1, diphenyl guanidine 1, stearic acid 1 and sulfur 1.5. After 40 mins. of milling, the homogeneous stock was cured for 30 minutes at 160°. Extraction was carried out with n-hexane at 65° for 24 hours.

III. Results and Discussion

1. Glass Transition Temperature Studies

The glass transition temperatures of a series of poly(isobutylene-co-β-pinene) samples of varying compositions have been determined. Table 1 shows the results.

Even a cursory examination of the data indicates increasing Tg with increasing β-pinene content reflecting the influence of the "hard" β-pinene repeat unit in the copolymer. The Tg's have been correlated quantitatively with copolymer composition using the following Eq. (2):

$$Tg = [Tg_1 + (K\,Tg_2 - Tg_1)V_2]/[1 + (K-1)V_2] \tag{1}$$

where Tg, Tg_1, Tg_2 are the glass transition temperatures of the copolymer, homopolymer 1 and homopolymer 2, respectively, and K a constant depending on the copolymer system. V_2, the volume fraction, was calculated by assuming that the volumes of each carbon atoms are constant and additive.

Equation (1) (in the form shown or in an equivalent form) has often been used to correlate the composition of random copolymers with Tg (2).

Table 1. Glass transition temperature of poly(isobutylene-co-β-pinene)

Sample[a]	β-pinene (mole%)	β-pinene (volume% or V_2)	$M_v \times 10^{-3}$	$Tg \pm 4$ °C
Polyisobutylene	0	0	1,000	−70
3	3	7	650	−65
4	10	22	92	−53
5	18	35	27	−28
6	41	63	10	+12

[a] Samples described in Table 1 of the first paper of this series.

It is difficult to fit our data to this equation because K and Tg_2 are not known. Also, valid composition and Tg correlation can only be derived with high molecular weight materials in the range where Tg is insensitive to molecular weights. While the molecular weights of our copolymers containing up to 35 volume % (18 mole %) β-pinene can definitely be considered high and even those containing 63 volume % (41 mole %) β-pinene are probably acceptable for this purpose, the molecular weights of our pure poly(β-pinene) samples were too low to be used directly in the equation. Repeated efforts have been made to dermine the Tg of our poly(β-pinene) samples, however, due to the low molecular weight of the polymer, unsuitable physical arrangement of our DSC instrument, and the overall shape of the DSC trace we were unable to obtain a reproducible value.

The Tg of pure poly(β-pinene) was estimated by Wood's method (2), by plotting the Tg of our copolymers versus composition (volume fraction of β-pinene in the copolymer, V_2) and extrapolating to 100% poly(β-pinene). Figure 1 shows the data:

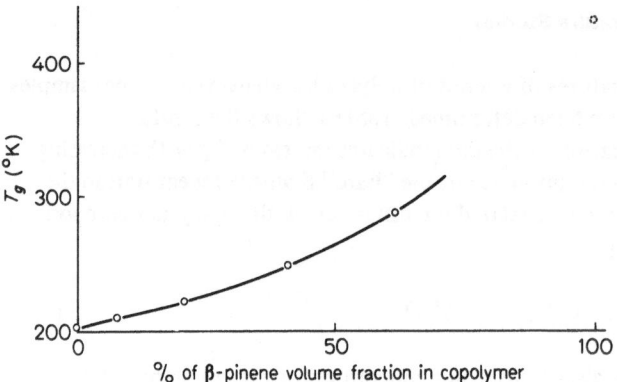

Fig. 1. Glass transition temperature dependence on poly(isobutylene-co-β-Pinene) Composition

the estimated Tg of poly(β-pinene), 165 °C, is indicated by the dotted circle. The constant K was subsequently calculated to be 0.33 for this system.

Thus by the use of Wood's approximation we were able to construct a complete Tg/composition curve for the poly(isobutylene-co-β-pinene) system. In view of the experimental difficulties in obtaining the Tg for poly(β-pinene), the accuracy of the plot is questionable beyond $V_2 = 63$ volume% $i.e.$, above 63 volume% β-pinene in the copolymer. However, in the range from 0 to 63 volume% β-pinene, the curve is considered to be accurate and, importantly, reveals that useful rubbery poly(isobutylene-co-β-pinene) could be made with up to about 28 volume% β-pinene, $i.e.$, to a Tg of $-40\,^\circ$C. In agreement with our spectroscopic and reactivity ratio studies, the (admittedly somewhat limited) applicability of Eq. (1) developed for random copolymers also suggests a statistically random copolymer structure for our poly(isobutylene-co-β-pinene) products.

2. Sulfur Vulcanization of Rubbery Poly(isobutylene-co-β-Pinene)

The results of a series of runs carried out to obtain amounts of copolymers sufficient for physical property characterization are summarized in Table 1 of the first paper of this series. It is evident that depending on the relative amount of isobutylene and β-pinene units in the copolymer, poly(isobutylene-co-β-pinene)s are rubbery, leathery, or glassy materials.

Copolymers containing 3 mole% unsaturation were found to form insoluble gels during storage for about two months. Indeed, this network formation was the first indication for successful copolymerization. Gel formation can be avoided by the use of an antioxidant.

As shown in Table 1, poly(isobutylene-co-β-pinenes) containing up to ~ 10 mole% β-pinene are rubbery materials ($Tg < -53^\circ$). In view of the structural similarity between our copolymer and butyl rubber, we decided to use conventional butyl rubber cure conditions (3) to prepare our vulcanizates (see Experimental). Figure 2 shows the curing rate of two copolymers containing 3 and 10 mole% β-pinene units respectively, as determined by a Monsanto Rheometer at 160 $^\circ$C (320 $^\circ$F). As expected, the curing rate of the copolymer containing 10 mole% unsaturation is several times higher than that of the material with 3 mole% unsaturation level.

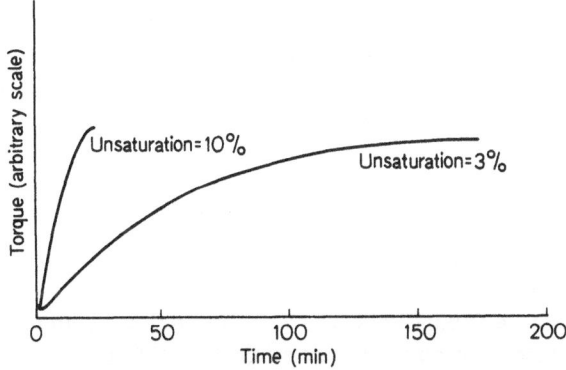

Fig. 2. The curing rate of poly(isobutylene-co-β-pinene). Monsanto rheograph, 160°

Some representative mechanical properties of poly(isobutylene-*co*-β-pinene) vulcanizates are shown in Fig. 3. The definition of optimum cure condition for rubbery poly(isobutylene-*co*-β-pinene) is beyond the scope of this research.

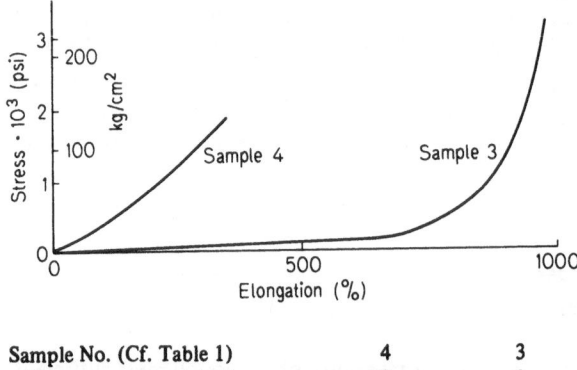

Fig. 3. Some mechanical properties of poly(isobutylene-*co*-β-pinene) vulcanizates

Sample No. (Cf. Table 1)	4	3
β-Pinene content, mole %	10	3
HAF black, PHR	40	0
Tensile strength, psi	2,100	3,500
kg/cm^2	148	246
300% modulus, psi	1,700	110
kg/cm^2	120	7.7
Elongation, %	350	950
Hardness, Shore "A"	80	28
Permanent Set, %	–	12

Testing conditions in Experimental Part

The mechanical properties of vulcanizates of isobutylene-rich copolymers are similar to that of commercial butyl rubbers. The stretched rubber showed birefringience under polarizing lenses, indicating crystallization on stress, a phenomenon commonly observed during stress stiffening. These vulcanizates are very soft rubbers with high ultimate tensile strength when compounded without reinforcing fillers. The moduli of these materials increases substantially when compounded with carbon black filler.

3. Ozone Resistance of Poly(isobutylene-*co*-β-Pinene) Vulcanizates

Poly(isobutylene-*co*-β-pinene) vulcanizates are ozone resistant. Even though ozone may cleave the double bond in the β-pinene unit, the molecular weight of the macromolecule will not decrease because of the "protected" nature of the unsaturation:

To determine the ozone resistance of poly(isobutylene-*co*-β-pinene), cut dumbbells of poly(isobutylene-*co*-β-pinene) vulcanizates were stretched to 100% extension and placed in a glass chamber at room temperature with constant ozone flow. The ozone concentration was kept at 2.78 mg/l or 1,400 ppm which is about 10,000 times the ozone concentration in the atmosphere. We also tested samples of natural rubber, butyl rubber and EPDM rubber vulcanizates together under the same condition for comparison. Results are summarized in Table 2.

Table 2. The effect of ozone exposure on various rubber vulcanizates. (ozone concentration: 1,400 ppm)

Vulcanization system	Ozone exposure time, hours	
Poly(isobutylene-co-β-pinene)[a]	5	No observable changes stress/strain response remains unaffected, see Fig. 4
Butyl rubber[b]	2	Broken, surface became oily, indicating surface degradation
EPDM rubber[c]	5	No observable changes
Natural rubber[d]	Few minutes	Crumbled

[a] Curing recipe, parts: Polymer 3 (see Table 1 of first paper of this series) 100, Zinc Oxide 5, Tetramethylthiuram Disulfide 1, Benzothiazyl Disulfide 1, Stearic Acid 1, Sulfur 1.25, cured 35 minutes at 160 °C.

[b] Curing recipe, parts: Polymer F8NS (Copolymer Co., Sarnia, Ontario) 100, Zinc Oxide 5, Tetramethylthiuram Disulfide 1, Benzothiazyl Disulfide 1, Stearic Acid 1, Sulfur 1.25, cured 35 minutes at 160 °C.

[c] Curing recipe, parts: Epsyn 40A (Copolymer Co., Sarnia, Ontario) 100, Zinc Oxide 5, Santocure (Monsanto) 1, Diphenyl Guanidine 0.5, Stearic Acid 1, Sulfur 1.25, cured 35 minutes at 160 °C.

[d] Curing recipe, parts: Smoked Sheet 100, Zinc Oxide 5, Tetramethylthiuram Disulfide 0.1, Benzothiazyl Disulfide 1, Stearic Acid 2, Sulfur 2.5, cured 20 minutes at 143 °C.

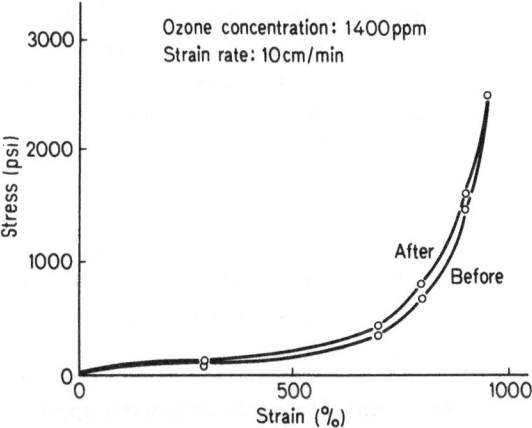

Fig. 4. Stress-strain response of poly(isobutylene-co-β-pinene) vulcanizate before and after ozone exposure

The stress-strain response of a poly(isobutylene-*co*-β-pinene) vulcanizate, containing 3% β-pinene, was determined before and after ozone exposure and is shown in Fig. 4. The finding that the stress-strain response was virtually unaffected by ozone confirms the "protected" nature of unsaturation in poly(isobutylene-*co*-β-pinene).

4. Covulcanization Studies

Polymer of high molecular weight are in the main incompatible, but it is often advantageous to compound two polymers in order to obtain a combination of new physical properties which either polymer alone cannot have. For example, EPDM rubber is incompatible with diene rubbers such as polyisoprene, polybutadiene, SBR etc., yet it has been covulcanized with those rubbers to provide useful materials (4). In this work, we have found that poly(isobutylene-*co*-β-pinene) with 10 mole% unsaturation can also be cocured with a high cis-1,4-polybutadiene rubber. Covulcanization has been demonstrated by an experiment in which poly(isobutylene-*co*-β-pinene) plus cis-1,4-polybutadiene were mixed on a laboratory mill for 40 minutes, cured (30 minutes at 160°) and the vulcanizate extracted with n-hexane (24 hours at 65°). The fact that the amount of extractables was 2% indicates a measure of co-vulcanization. The covulcanizate exhibited a tensile strength of 175 kg/cm^2 and elongation at break of 300%.

IV. Summary and Conclusions

Some physical properties of random poly(isobutylene-*co*-β-pinene) have been investigated. These copolymers are rubbery, leathery or glassy materials depending on the relative amount of isobutylene and β-pinene units in the chain. Tg's determined by differential scanning calorimetry range from −65° for copolymers containing 3 mole% β-pinene to 12° for 41 mole% β-pinene. The general trend of the Tg/composition curve is in accord with that exhibited by statistical copolymers. (Fig. 1).

Copolymers containing up to ∼10 mole% β-pinene are rubbery (Tg: −53°) and can be sulfur vulcanized. The non-filled vulcanizates are very soft rubbers (*e.g.*, 300% modulus: 7.7 kg/cm^2 or 110 psi) with high ultimate tensile strength (*e.g.*, tensile strength: 246 kg/cm^2 or 3,500 psi). The high tensile strength is attributed to stress induced crystallization.

Due to the "protected" nature of internal double bonds, poly(isobutylene-*co*-β-pinene) is ozone resistant. For example, the stress-strain response of a vulcanizate containing 3 mole% β-pinene showed virtually no change after ozone exposure (Fig. 4 and Table 2).

Acknowledgements. We are grateful to the National Science Foundation for financial support of this research.

V. References

1. Wadelin, C. W.: Anal. Chem. **29**, 411 (1957).
2. Wood, L. A.: J. Polymer Sci. **28**, 319 (1958).
3. The vanderbilt rubber handbook, (ed. G. G. Winspear). New York: Vanderbilt Publishers 1968.
4. OMahaney, J. F., Jr.: Rubber Age **120**, 37 (1970).

Received October 26, 1975

Ring-Chain Equilibria and the Conformations of Polymer Chains

Joseph A. Semlyen

Department of Chemistry, University of York, Heslington, York YO1 5DD

Table of Contents

I. Introduction

A. Ring Molecules in Polymeric Systems

Cyclic or ring molecules occur in a wide range of polymers, including linear and non-linear polymers prepared by condensation or by ring-to-chain equilibration reactions. Cyclics also occur in many biological macromolecules, sometimes in the form of loops as in certain proteins and nucleic acids.

The presence of ring molecules has to be taken into account in the manufacture of some commercially-important polymers including linear poly(dimethylsiloxane), nylon-6 and poly(ethylene terephthalate). Indeed much of the research carried out on cyclic concentrations in such polymers has been by workers in Industry, who have been concerned to acquire detailed knowledge of the cyclic contents of their products.

The principal aim of this review is to summarize experimental and theoretical studies of the cyclic populations of polymers *prepared by ring-chain equilibration reactions.* In particular, to show how individual cyclic concentrations (determined experimentally) may be related directly to the statistical conformations of the corresponding open chain molecules. The polymers range from the structurally simple (such as polycatenasulphur) to the structurally more complicated [such as block copolymers of polystyrene and poly(dimethylsiloxane)]. The investigations discussed here are confined as far as possible to chemical systems where there is a simple thermodynamic equilibrium between ring and chain molecules, the latter having a most probable distribution of chain lengths. However, it is believed that at least some of the data relating to the cyclic populations of linear polymeric equilibrates presented here are relevant to the more complicated types of system discussed by Stepto (*1*) in his review. These latter systems include non-linear polymers, as well as linear polymers prepared under kinetic control. They have been treated theoretically [see, for example, Refs. (*2–8*)] and investigated experimentally [see, for example, Refs. (*9–12*)]. Another example of a complex intramolecular cyclization reaction in a polymeric system that has been described in some detail is that between reactive or catalytic groups spaced along polymeric chains [see, for example, Refs. (*13–15*)].

B. Equilibrium Cyclic Concentration Method for Studying the Conformations of Chain Molecules

Investigations of the cyclic populations of simple ring-chain equilibrates such as those set up in polysiloxane and polyester systems have at least two major advantages over the study of ring formation in more complex systems, such as kinetically-controlled non-linear polymerisations.

The first advantage is the ease with which the cyclic and linear polymers can be separated and analysed. Much use is made of the techniques of gas-liquid chromatography (g.l.c.) for volatile molecules, and gel permeation chromatography (g.p.c.) for the less volatile molecules. As examples of the power of these methods, it is noted that cyclic dimethylsiloxanes $[(CH_3)_2SiO]_x$, with up to a hundred skeletal bonds, can be completely separated from one another by g.l.c. using a commercial instrument (16). Also, cyclic ethylene terephthalates $(CO \cdot C_6H_4 \cdot CO \cdot O \cdot CH_2 \cdot CH_2 \cdot O)_x$ with $x = 3–9$ can be separated without prior fractionation using a g.p.c. instrument (17). Normally both chromatographic techniques are used for a particular system and the uncertainties in the values for cyclic concentrations in quenched ring-chain equilibrates are typically ±5–10% for the smaller cyclics and ±10–30% for cyclics with up to a hundred skeletal bonds. However, if sufficient care is taken, greater precision can be obtained for the larger cyclics in favourable cases such as poly(dimethylsiloxane), as was suggested in the pioneering work of Brown and Slusarczuk (18) and later confirmed by Wright (16, 19, 20).

The second advantage of studying cyclic concentrations in linear, equilibrated systems is that the Jacobson and Stockmayer theory (21) can be used to relate macrocyclic equilibrium concentrations directly to the average conformations of the corresponding open chain molecules. In fact, the measurement of equilibrium cyclic concentrations provides a powerful new technique for characterising the average conformations of chain molecules. The equilibrium cyclic concentration method is applicable to a wide range of polymeric systems as will be apparent from the investigations reviewed here. Furthermore, the method may be used to study the conformations of chain molecules over a range of chain lengths from only a few to as many as several hundred skeletal bonds. These measurements may be made in a variety of media including polymeric melts, polymers in solution and polymers in the solid state.

C. General Methods for Studying the Conformations of Polymer Chains

Information relating to the average conformations of chain molecules in dilute solution can be obtained using a variety of experimental techniques. These include light scattering, viscometric, sedimentation, diffusion, dipole moment, dielectric relaxation, electric birefringence, electric and circular dichroism and nuclear magnetic resonance methods [see, for example, Refs. (22–27)]. Measurements of the stress-temperature coefficients of cross-linked networks can be used to study the effect of temperature on the average dimensions of chains in bulk polymers (28). A relatively new technique, small angle neutron scattering, is being increasingly applied to investigate the conformations of chains in amorphous polymers and polymeric melts as well as in solution (29–33). It is more effective for polymers than the small angle X-ray scattering method employing molecules tagged at their ends with suitable scattering groups (34, 35).

The widely applied light scattering and viscometric methods give a measure of the size of macromolecular coils in solution through their radii of gyration (26, 27). By contrast, the equilibrium cyclic concentration method, discussed here, gives a measure of the probabilities of intramolecular cyclization of chain molecules over a range of chain lengths. These probabilities will be related directly to the radii of gyration for rings corresponding to long random-coil chains, but, in general, they will not be so for the smaller rings (see later).

D. Theoretical Descriptions of the Conformations of Chain Molecules

In order to establish the equilibrium cyclic concentration method for studying chain conformations, the Jacobson and Stockmayer cyclization theory (*21*) was modified (*36*) to make it applicable to real chain molecules. In particular, the mathematical methods developed by Flory and his co-workers (*27, 37, 38*) were used to calculate average values for the dimensions of random-coil chains of finite length, making use of realistic rotational isomeric state models to describe their conformational statistics. Such models were originally developed by Volkenstein (*39*), by Birshtein and Ptitsyn (*40*) and by others (*41–43*). Existing models were modified and other models set up during an extended research programme by Flory and his co-workers (*27, 44*). In brief, rotational isomeric state models assume that each skeletal bond of a polymer chain is confined to one of a discrete number of rotational isomeric positions, usually corresponding to minima in the underlying torsional rotational potentials. The average dimensions, or other statistical properties of the polymeric chains, are then calculated, allowing for the relative probabilities of the rotational states and making full allowance for the mutual interdependence of adjacent pairs of bond rotational states. Parameters required for the models are normally obtained from molecular structure data and from experimental information relating to the statistical conformations of the polymeric chains. Further references to rotational isomeric state models will be made later.

II. Jacobson and Stockmayer Cyclization Theory

A. Ring-Chain Equilibration Reactions

This review is concerned primarily with the distribution of cyclic species in polymeric systems where there are thermodynamic equilibria between ring and chain molecules. Such equilibria may be represented as follows

$$-M_y- \; \rightleftharpoons \; -M_{y-x}- \; + \; M_x \tag{1}$$

where $-M_y-$ and $-M_{y-x}-$ represent linear molecules and M represents one monomer unit. The terminal groups of the linear molecules depend, in general, on the chemical nature of the catalyst used to attain ring-chain equilibrium. Their nature does not affect the positions of the equilibria and so they do not enter into any theoretical considerations relating to equilibrium cyclic concentrations.

For convenience, the cyclic populations of ring-chain equilibrates are expressed in terms of the molar cyclization equilibrium constants K_x for the individual x-meric ring molecules M_x, thus

$$K_x = \frac{[-M_{y-x}-][M_x]}{[-M_y-]} \; . \tag{2}$$

Provided that there is a most probable distribution (a Flory distribution) of chain
lengths in the linear part of the equilibrate, Eq. (2) reduces to

$$K_x = [M_x]/p^x \tag{3}$$

where p is the extent of reaction of functional groups in the chain polymer. For
most of the systems to be discussed here, p is close to unity and so the K_x values for
all but the largest cyclics are close to the values of the molar concentrations of the
cyclics, usually expressed in mol dm^{-3}. In practice, values of p are found by viscometric, osmometric, light scattering or g.p.c. methods; or, in favourable cases, by end-
group analysis. The values of p found are combined with the measured molar concentrations of cyclics in Eq. (3) in order to yield experimental K_x values.

B. Cyclization of Chains Obeying Gaussian Statistics

Kuhn (45, 46) was the first to establish that real polymer chains in random-coil conformations obey the following Gaussian expression for the density $W_x(r)$ of their
end-to-end vectors r

$$W_x(r) = (3/2 \pi <r_x^2>)^{3/2} \exp(-3 r^2/2 <r_x^2>) \tag{4}$$

The applicability of this relationship to chain molecules depends on their length and
flexibility and on the value of the scalar length r. The relationship applies closely to
very long chains in random-coil conformations at moderate extensions, i.e., when
the values of r are relatively small (46). From Eq. (4) it follows that the density of
the distribution $W_x(r)$ in the region $r = 0$ for long chain molecules obeying Gaussian
statistics is

$$W_x(0) = (3/2 \pi <r_x^2>)^{3/2} \tag{5}$$

where $<r_x^2>$ represents the mean-square end-to-end distances of x-meric chains and
the units of $W_x(0)$ are molecules dm^{-3}.

C. Theoretical Expression for Molar Cyclization Equilibrium Constants

In their classic paper published in 1950, Jacobson and Stockmayer (21) presented a
theory giving a precise expression for the concentrations of macrocyclics M_x in ring-
chain equilibrates. The molar cyclization equilibrium constants K_x were given by

$$K_x = (3/2 \pi <r_x^2>)^{3/2} (1/N_A \sigma_{Rx}). \tag{6}$$

The K_x values are in mol dm^{-3}, N_A is the Avogadro constant and σ_{Rx} is a symmetry number or more generally the number of skeletal bonds of an x-meric ring that can open in the reverse reaction of Eq. (1). Values of either x or $2x$ are assigned to σ_{Rx} in all the systems to be discussed in Section 3.

Now Eq. (6) applies to cyclics corresponding to chains of sufficient length and flexibility to obey Gaussian statistics so that $W_x(0)$ is given by Eq. (5). This will not be so for chains corresponding to smaller rings and a more general expression for K_x is required (36)

$$K_x = W'_x(0)/N_A \, \sigma_{Rx} \tag{7}$$

where $W'_x(0)$ represents the density of end-to-end vectors corresponding to the close approach of chain ends, the prime indicating that the density is non-Gaussian and takes full account of the relative orientations of terminal bonds of chains undergoing intramolecular cyclization as opposed to intermolecular condensation reactions (36).

Finally, little attention will be paid to the smallest cyclics present in ring-chain equilibrates. For such cyclics, ΔH_x^0 for Eq. (1) will differ appreciably from zero and their K_x values will be sensitive to the temperature of equilibration. In general, attention will be confined to larger cyclics for which the enthalpy changes for the forward and back reactions of Eq. (1) can be taken to equal zero.

D. Calculation of Theoretical Molar Cyclization Equilibrium Constants

The Jacobson and Stockmayer theory (21) yields a simple theoretical expression for the molar cyclization equilibrium constants for macrocyclics in ring-chain equilibrates. K_x values can be calculated directly by Eq. (6) provided the corresponding $<r_x^2>$ values are known. Conversely, $<r_x^2>$ values can be obtained by simply measuring K_x values.

Excluded volume effects will not be present in systems where ring-chain equilibration is carried out in the undiluted polymer or in concentrated polymer solutions (26); furthermore, they have been found to be negligible for chains with less than about a hundred skeletal bonds, when the chains are in good solvents [see Refs. (18–20)]. Hence, in general, $<r_x^2>$ values can be identified with $<r_x^2>_0$ values characteristic of chains in θ-solvents (or Flory solvents). The characteristic ratios $C_\infty = (<r_x^2>_0/nl^2)_\infty$ of the mean-square end-to-end distances of chains containing n skeletal bonds each of length l (taken in the limit $n \to \infty$) have been measured for many of the polymers to be discussed here. Furthermore, realistic rotational isomeric state models for these polymers have been set up. Hence, the powerful mathematical methods of Flory and Jernigan (37, 38) can be used to compute exact values for $<r_x^2>_0$ over the complete range $1 < x < \infty$. Such values can be substituted into Eq. (6) to provide theoretical K_x values, which can be compared with those measured experimentally. Within this theoretical framework, detailed considerations of the cyclic populations of some real polymeric equilibrates are undertaken in the following section.

III. Cyclic Populations of Polymeric Ring-Chain Equilibrates

A. Polysiloxanes

1. Studies of Cyclic Methylsiloxanes

The cyclic populations of a range of equilibrated polymeric methyl siloxanes $(R(CH_3)SiO)_x$ where R is hydrogen, methyl, ethyl, n-propyl, 3,3,3-trifluoropropyl and phenyl have all been studied in recent years (*16, 18–20, 47, 48*). Of these, the poly(dimethylsiloxane) system has received the most attention. It is the most important commercial silicone and it has proved to be more amenable to detailed experimental study than other silicones including those listed above.

The first extensive investigation of ring-chain equilibration in poly(dimethyl-siloxane) polymers was by Scott (*49*), this was followed by the studies of Hartung and Camiolo (*50*), of Carmichael and his co-workers (*51–54*) and of Brown and Slusarczuk (*18*). These workers established that poly(dimethylsiloxane) can be equilibrated in the undiluted polymer as well as in solution. Equilibrium between ring and chain molecules can be set up using a range of catalysts and the equilibrium can be quenched prior to analysis. Brown and Slusarczuk (*18*) demonstrated that the powerful techniques of g.l.c. and g.p.c. can be used to measure the molar cyclization equilibrium constants K_x over the range of $x = 3$ to $x = ca$. 300, provided that molecular distillation and solvent extraction are used to separate the cyclics, as far as possible, from chain molecules and from each other.

2. Chromatographic Techniques

The power of the modern techniques of g.l.c. and g.p.c. for the characterisation of the cyclic populations of polymers is illustrated for some polymethylsiloxane systems in Figs. 1–3. These chromatograms are typical of those obtained over virtually the whole range of systems studied and discussed in this review. Figure 1 shows a gel permeation chromatogram of a quenched, undiluted ring-chain equilibrate of poly-(phenylmethylsiloxane). It shows small cyclics $(C_6H_5(CH_3)SiO)_x$ with $x = 3, 4,$ 5, . . . separated from linear chains, with an indication that macrocyclics are present in the intervening region (*48*). Figure 2 shows a typical gas-liquid chromatogram of a cyclic fraction obtained by fractional distillation of rings from a ring-chain equilibrate of poly(ethylmethylsiloxane), cyclics $(CH_3CH_2(CH_3)SiO)_x$ with $x = 12–17$ make up nearly all the sample. Finally, Fig. 3 shows the gel permeation chromatogram of macrocyclic dimethylsiloxanes, showing a mixture of cyclics $((CH_3)_2SiO)_x$ with about thirty to over five hundred skeletal bonds (*16, 20*).

3. Cyclic Dimethylsiloxanes

The first application of the equilibrium cyclic concentration method to measure precisely the average conformations of chain molecules in a polymer was for poly-(dimethylsiloxane) (*16, 18–20, 36*). In Fig. 4, K_x values for a bulk equilibrate at 383 K (*19*) are compared with the corresponding values for a solution equilibrate (in toluene at a siloxane concentration of 224 g dm^{-3}) [Refs. (*18, 20*)] and with theo-

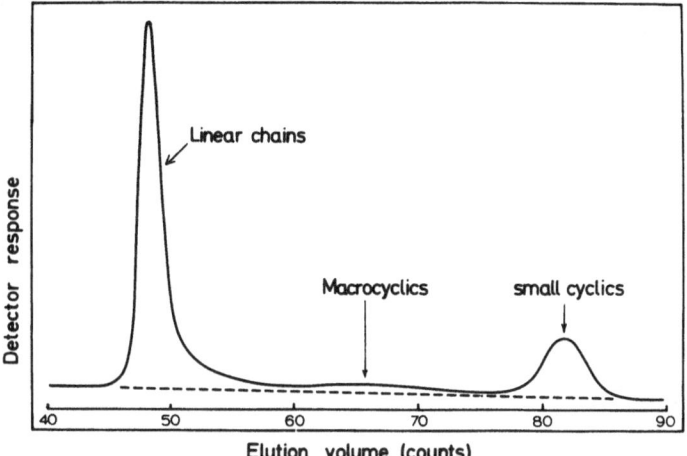

Fig. 1. Gel permeation chromatogram (g.p.c.) of an undiluted equilibrate of poly(phenylmethyl-siloxane) showing regions corresponding to small cyclics, macrocyclics and linear polymer

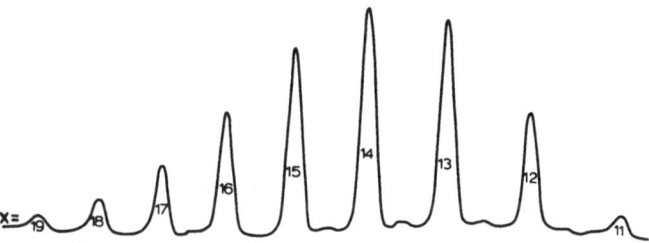

Fig. 2. Gas-liquid chromatogram (g.l.c.) of a cyclic fraction from a ring-chain equilibrate of poly-(ethylmethylsiloxane) containing cyclics $(CH_3CH_2(CH_3)SiO)_x$ with $x = 11-19$

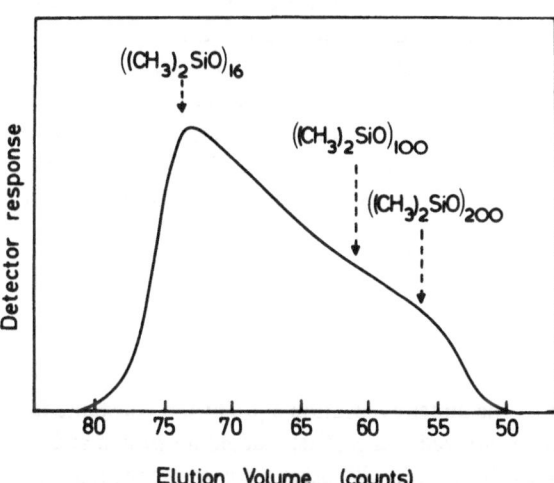

Fig. 3. Gel permeation chroma-togram (g.p.c.) of macrocyclic dimethylsiloxanes $((CH_3)_2 SiO)_x$ containing about thirty to over five hundred skeletal bonds

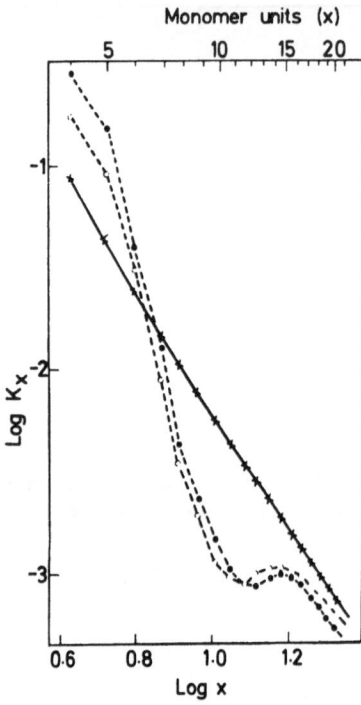

Fig. 4. Molar cyclization equilibrium constants K_x for cyclics $((CH_3)_2SiO)_x$ at 383 K for a bulk equilibrate (o) are compared with a solution equilibrate (•) and with values calculated (x) as described in the text. K_x values are in mol dm^{-3}

retical values calculated (*36*) by the Jacobson-Stockmayer expression [Eq. (6)] with $\sigma_{Rx} = 2x$. Values of $< r_x^2 >$ required to calculate the K_x values were computed by the methods of Flory and Jernigan (*37, 38*) by assuming that chains were unperturbed by excluded volume effects, and using the rotational isomeric state model for poly(dimethylsiloxane) set up by Flory, Crescenzi and Mark (*55*) on the basis of their own experimental data relating to the conformations of polymeric dimethylsiloxanes (*56, 57*). As shown in Fig. 4, good agreement between experiment and theory is obtained for cyclics with more than thirty skeletal bonds, and the results show that dimethylsiloxane chains of intermediate length adopt similar conformations in the undiluted polymer, in θ-solvents and in a solution consisting mainly of toluene and small siloxane rings and containing less than 1% linear polymer. These results show that chains adopt random-coil conformations obeying Gaussian statistics in the different media. They contradict the theories of Kargin (*58*) Hosemann (*59*), Pechhold (*60*), and Yeh (*61*), which predict degrees of order in all amorphous polymers including undiluted poly(dimethylsiloxane).

4. Macrocyclic Dimethylsiloxanes

Extensions of the measurements of the molar cyclization equilibrium constants for dimethylsiloxane cyclics into the macrocyclic range, first studied by Brown and Slusarczuk (*18*), have been made by Wright (*16, 20*) using g.p.c. as the main analyt-

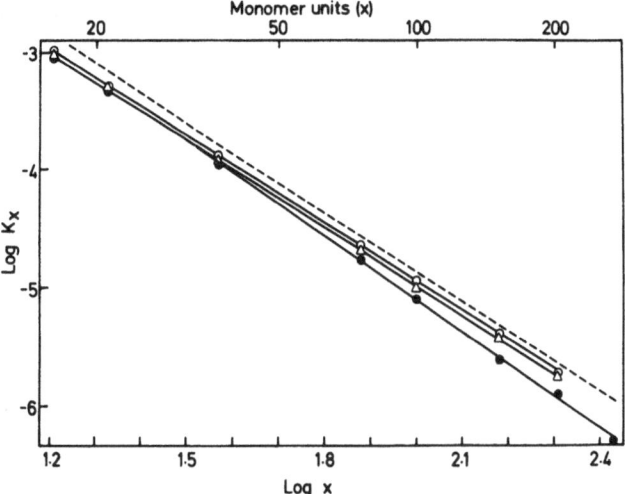

Fig. 5. Molar cyclization equilibrium constants K_x for cyclics $((CH_3)_2SiO)_x$ in an undiluted equilibrate at 383 K (o) are compared with values in diglyme at 333 K ($\triangle$), in toluene at 383 K ($\bullet$) and with calculated values (—––) (see text). K_x values are in mol dm^{-3}

ical technique. Figure 5 shows experimental $\log K_x$ values plotted against $\log x$ for an undiluted equilibrate at 383 K, a ring-chain equilibrate carried out in a poor solvent (diethylene glycol dimethylether, otherwise known as diglyme) at 333 K and Brown and Slusarczuk's ring-chain equilibrate in toluene at 383 K (for which values extend to cyclics with more than five hundred skeletal bonds). These experimental values are compared with theoretical values calculated as described above (36). Values of the ratio $C_x = <r_x^2>/2xl^2$ [where the length of a siloxane bond $l = 0.164$ nm (55)] can be deduced directly from the data shown in Fig. 5. For the undiluted polymer $C_{100} = 7.1$ and for the diglyme system $C_{100} = 7.6$, whereas the Flory, Crescenzi and Mark (55) rotational isomeric state model predicts $C_{100} = 6.4$ at 383 K.

Wright's results for the ring-chain equilibrate carried out in toluene solution, as described by Brown and Slusarczuk (18) are in broad agreement with the findings of these authors. The gradient in the plot of $\log K_x$ against $\log x$ over the range $x = 16$ to $x = ca.$ 40 is found to be -2.50 by both Wright and by Brown and Slusarczuk. However for $x = 40$ to $x = ca.$ 200, Wright finds a mean gradient of -2.69. This is less steep than the value of $-2.86(\pm5\%)$ reported by Brown and Slusarczuk and it is also less than the maximum gradient predicted by current theories of the excluded volume effect (62–64). [A theoretical maximum of -2.80 has been predicted (62)]. Wright's results lead to the conclusion that dimethylsiloxane chains of intermediate length are considerably expanded by excluded volume effects in dilute solution in a good solvent, as had previously been suggested by Brown and Slusarczuk. The magnitude of Flory's (26) expansion factor α obtained by Wright for the toluene system at $x = 100$ is $\alpha = 1.08$. The detailed and extensive work of Wright, based on the publication of Brown and Slusarczuk, gives definitive information relating to the average conformations of chain molecules (with chain lengths intermediate between those of oligomers and polymers) in a variety of environments.

The experimental techniques used at this molecular weight range should be applicable to a number of polymer systems including some of those mentioned in this section.

5. Effect of Substituent Groups on the Conformations of Siloxane Chains (47, 48)

An understanding of the thermodynamics of ring-chain equilibration reactions is important in Inorganic Chemistry where such reactions abound (see, for example, Allcock (65), Gee (66), Gimblett (67), and Moedritzer (68)). Two such reactions are found (i) in polyphosphazene chemistry (69) where a polymerisation-depolymerisation equilibrium can be established (70); and (ii) in silicon-sulphur chemistry, where the analogue of the poly(dimethylsiloxane) system has been studied (71) and shown to contain cyclics such as the dimer ($(CH_3)_2 SiS)_2$ and the trimer ($(CH_3)_2 SiS)_3$, as well as small amounts of chain molecules of limited chain length.

It was with such considerations in mind that a study of the effect of substituent groups R on the positions of ring-chain equilibria of a series of substituted polymethylsiloxanes ($R(CH_3)SiO)_x$ was undertaken (47, 48). Cyclics ($R(CH_3)SiO)_x$ in quenched ring-chain equilibrates were analysed for $x = 4-15$ (R = H) [see also Refs. (72, 73)], $x = 4-20$ (R = CH_3CH_2), $x = 4-8$ (R = $CH_3CH_2CH_2$), $x = 4-20$ (R = $CF_3CH_2CH_2$) [see also Ref. (74)] and $x = 3-50$ (R = C_6H_5). Equilibration temperatures of 383 K were used for all systems except R = H, which was studied at 273 K. This

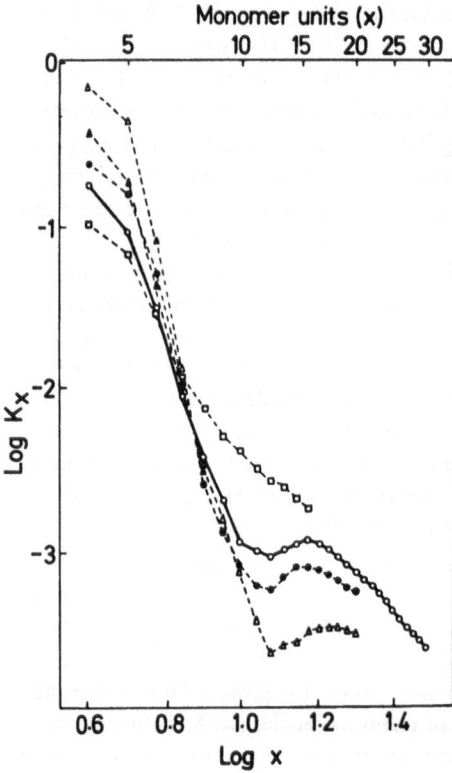

Fig. 6. Experimental molar cyclization equilibrium constants K_x for cyclics ($R(CH_3)SiO)_x$ in undiluted equilibrates at 273 K (R = H) and 383 K (R = CH_3, CH_3CH_2, $CH_3CH_2CH_2$, $CF_3CH_2CH_2$). K_x values are denoted by: □ for R = H; ○ for R = CH_3; ● for R = CH_3CH_2; ▲ for R = $CH_3CH_2CH_2$; △ for R = $CF_3CH_2CH_2$. K_x values are in mol dm^{-3}

temperature difference of 110 K is not expected to have an appreciable effect on the cyclic concentrations (47). Limits of accuracy of ±10% were placed on all the K_x values obtained, for the ranges of x specified above. These values are plotted as log K_x against log x in Fig. 6, where they are compared with the corresponding values for the dimethylsiloxanes. The K_x values for phenylmethylsiloxanes are not plotted as they are the same (within experimental error) as values for ethylmethylsiloxanes (48). This observation suggests that phenyl and ethyl groups have similar influences on the statistical conformations of siloxane chains. K_x values for the cyclic trimers $(R(CH_3)SiO)_3$ are omitted from consideration as they are low and temperature dependent. The results shown in Fig. 6 demonstrate that there is a striking correlation between the K_x values and the size of the substituent group R. The K_x values for the smallest, unstrained rings ($x = 4$ or 5) increase along the series $R = H < CH_3 < CH_3CH_2 < CH_3CH_2CH_2 < CF_3CH_2CH_2$, so that K_4 for $(H(CH_3)SiO)_4$ is lower than K_4 for $(CF_3CH_2CH_2(CH_3)SiO)_4$ by a factor of seven; by contrast, the K_x values for the larger cyclics decrease with increasing size of the group R, so that K_{12} for $(H(CH_3)SiO)_{12}$ is ten times larger than K_{12} for $(CF_3CH_2CH_2(CH_3)SiO)_{12}$. There are well defined minima in the log K_x versus log x plots in Fig. 6 for cyclics with $R = CH_3CH_2$ and $R = CF_3CH_2CH_2$ just as in the dimethylsiloxane system, but this is not the case for hydrogenmethylsiloxanes where there is only a point of inflection at $x = ca.$ 12. The hydrogenmethylsiloxanes and ethylmethylsiloxanes show the limiting slope of -2.5 predicted by the Jacobson and Stockmayer theory (21) [Eq. (6) with $\sigma_{Rx} = 2x$] in their log K_x versus log x plots at $x > ca.$ 12 and $x > ca.$ 15 respectively. Values of $<r_{18}^2>/36\,l^2$ for $(R(CH_3)SiO)_{18}$ chains containing 36 skeletal bonds (each of length 0.164 nm (55)) to $2xl^2$ can be deduced from K_x values in Fig. 6. They are 5.4 for $R = H$ at 273 K, 6.8 for $R = CH_3$ at 383 K and 8.6 for $R = CH_3CH_2$ at 383 K. These studies show the large effect of changing the group R on the position of ring-chain equilibria in substituted methylsiloxanes, and they demonstrate how detailed information relating to the statistical conformations of the siloxane chain molecules can be obtained.

6. Configurational Isomers of Phenylmethylsiloxane Cyclics

In connection with the studies of cyclic concentrations in polysiloxane ring-chain equilibrates described above, it is noted that there are configurational isomers for all the cyclics $(R(CH_3)SiO)_x$ because of the presence of asymmetric silicon atoms. There are two such configurational isomers for the cyclic trimers, and four for both the cyclic tetramers and the cyclic pentamers. Resolution of the cyclics $(R(CH_3)SiO)_x$ into separate isomers cannot readily be achieved when the substituent group R is an alkyl group, but all the configurational isomers of the phenylmethyl-siloxanes $(C_6H_5(CH_3)SiO)_x$ with $x = 3,4$ and 5 can be separated completely by g.l.c. alone. Following the studies of Lewis (75) Daudt and Hyde (76) Young (77) and Hickton (78) and their co-workers, the relative amounts of configurational isomers of cyclic phenylmethylsiloxanes $(C_6H_5(CH_3)SiO)_x$ with $x = 3-5$ were determined for undiluted and solution ring-chain equilibrates (48). The configurational isomers were found to be present in proportions close to those predicted to result from the random intramolecular cyclization of an atactic chain polymer. This suggests that stereospecific influences are absent from poly(phenylmethylsiloxane) ring-chain equilibrates, and there is no reason to believe that stereospecific influences are present in any of the other polysiloxanes discussed here.

7. Effect of Dilution on the Concentrations of Cyclic Siloxanes (47)

The total weight fraction of cyclics $(R(CH_3)SiO)_x$ with $x = 3$ to ∞ in undiluted high molecular weight polysiloxane ring-chain equilibrates are 12.5% for R = H, 18.3% for R = CH_3, 25.8% for R = CH_3CH_2, 30% for R = C_6H_5 and 82.7% for R = $CF_3CH_2CH_2$. One of the predictions of the Jacobson and Stockmayer theory (21) is that if a ring-chain equilibration is carried out in the presence of an inert diluent, the weight fraction of material in the form of cyclics should increase with increasing dilution up to a critical point beyond which linear polymer will be effectively absent. This follows from Eq. (6), which predicts that the molar concentrations of cyclics should remain approximately independent of dilution so that dilution converts chain to ring molecules. Fig. 7 shows the effect of dilution on the weight frac-

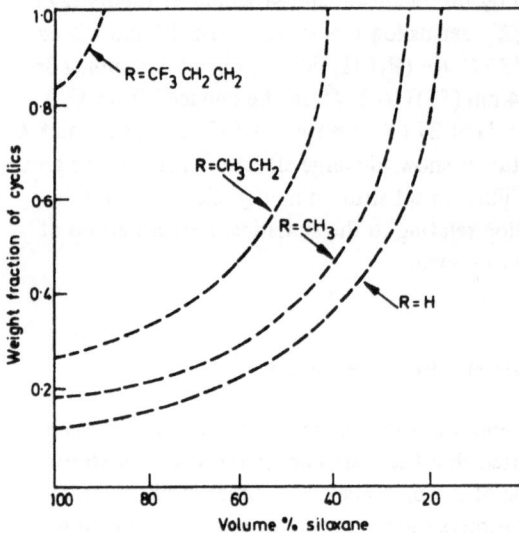

Fig. 7. Weight fractions of cyclics $(R(CH_3)SiO)_x$ in high molecular weight $(p \cong 1)$ polysiloxane equilibrates at 383 K (R = CH_3, CH_3CH_2, $CF_3CH_2CH_2$) and 273 K (R = H) as a function of the volume per cent siloxane in cyclohexanone (R = $CF_3CH_2CH_2$) and toluene (R = H, CH_3, CH_3CH_2)

tions of cyclics in four polymeric systems. The volume per cent of solvent required to attain the critical dilution point is *ca.* 10% cyclohexanone at 383 K when R = $CF_3CH_2CH_2$, *ca.* 60% toluene at 383 K when R = CH_3CH_2, *ca.* 75% toluene at 383 K when R = CH_3 and *ca.* 80% toluene at 273 K when R = H. In practice, however, the K_x values for the smaller cyclics in the poly(hydrogenmethylsiloxane), poly(dimethylsiloxane) and poly(ethylmethylsiloxane) equilibrates have been found to increase with dilution and the critical points are considerably lower than those predicted theoretically. In all these systems K_4 and K_5 increase by 50–100% over the dilution range to the critical dilution points, the K_6-K_8 values show marked though smaller increases (see Fig. 8 for the effects found for cyclics with R = H and R = CH_3CH_2).

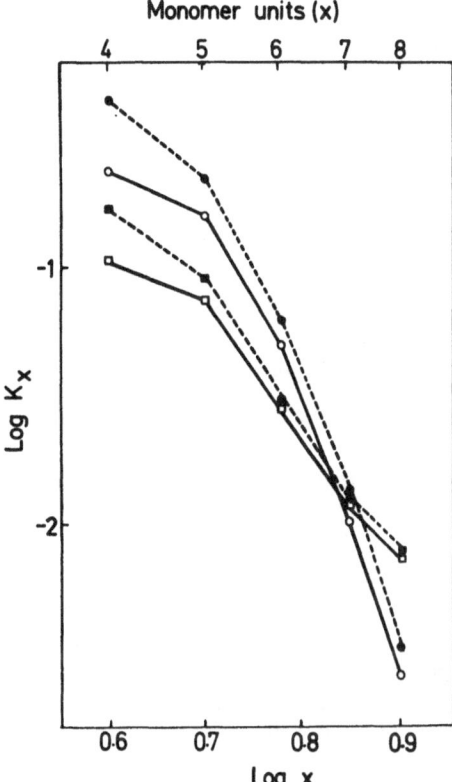

Fig. 8. Experimental molar cyclization equilibrium constants K_x for $(H(CH_3)SiO)_x$ at 273 K (denoted by □ and ■) and for $(CH_3CH_2(CH_3)SiO)_x$ at 383 K (denoted by ○ and ●). K_x values in undiluted equilibrates are denoted by □ and ○, and those measured in toluene solution close to the critical dilution points are denoted by ■ and ●. Units of K_x are mol dm^{-3}

Similar effects are found for R = C_6H_5 (*48*) but they cannot be observed for $CF_3CH_2CH_2$ over the limited dilution range (see Fig. 7). In all these systems the K_x values for the larger cyclics with $x > ca.$ 10 remain constant within experimental error [see, however, Section III. (A) 4 above for dilution effects on macrocyclic siloxanes].

An interesting contrast to these results is provided by the K_x values for small cyclics in poly(dimethylsiloxane) equilibrated in solution in diglyme (*16, 20*). In this case, the magnitude of the solvent effect is much smaller than in toluene; and, in fact, K_x values for $x = 4$ and $x \geqslant 6$ are identical within experimental error for an undiluted equilibrate and a diglyme equilibrate containing 134 g dm^{-3} cyclics and 55 g dm^{-3} chain polymer (*16, 20*).

These solvent effects can be interpreted semiquantitatively (*16*) in terms of polymer-solution thermodynamics by methods applied by Ivin, Leonard (*79, 80*) and others (*81, 82*) to simple monomer-polymer equilibria in solution. The poly(dimethylsiloxane) systems can be approximated as three-component systems with polymer, and solvent, and small cyclics constituting an effective monomer concentration (16). These results show that cyclic concentrations in polysiloxane ring-chain equilibrates are sensitive to the nature as well as the concentration of the solvent.

B. Polymeric Paraffin-Siloxanes (83)

The remarkable ease with which poly(dimethylsiloxane) enters into ring-chain equilibration reactions prompted the investigations described in this section, as well as those outlined in Section III.(C). Both investigations were concerned with the question as to whether it is possible to study the statistical conformations of chain polymers by the equilibrium cyclic concentration method, if the polymer itself cannot undergo ring-chain equilibration reactions. The polymers chosen for study were n-paraffinic and polystyrene chains. In each case, dimethylsiloxane linkages were incorporated into the molecules and ring-chain equilibration reactions were carried out.

The paraffin-siloxanes are a class of polymer containing methylene residues $+CH_2\rightarrow_n$ connected by siloxane linkages $-\overset{|}{Si}-O-\overset{|}{Si}-$. Polymers with $n = 2,3$ and 4 can be prepared from cyclic monomers first synthesised by Picolli and his co-workers (84), by Kumada and Habuchi (85) and by Sommer and Ansul (86). The polymer chosen for study was poly(2,2,7,7-tetramethyl-1-oxa-2,7-silacycloheptane). Samples were obtained from the corresponding cyclic monomer $((CH_3)_2 Si-(CH_2)_4-(CH_3)_2 Si-O)$ by ring-chain equilibration reactions using concentrated sulphuric acid (ca. 0.01% w/w) over a temperature range from 298 K to 423 K. Cyclics were separated from linear polymers, molecularly distilled into a number of sharp fractions and analysed by g.l.c. The linear polymers were investigated by g.p.c. Infra-red spectroscopy and proton nuclear magnetic resonance techniques were used to confirm that cyclic oligomers and linear polymers contained the repeat unit $(CH_3)_2 Si-(CH_2)_4-(CH_3)_2 Si-O$. Experimental K_x values for $x = 1-6$ were measured at 298, 333, 383 and 423 K. Apart from K_1, the K_x values were found to be independent of the temperature of equilibration. Experimental K_x values for the cyclics $((CH_3)_2 Si-(CH_2)_4-(CH_3)_2 Si-O)_x$ with $x = 1-6$ at 298 K are plotted as log K_x against log x in Fig. 9. They are seen to be in good agreement for $x \geqslant 3$ with theoretical values calculated by the Jacobson and Stockmayer theory using Eq. (6) with $\sigma_{Rx} = 2x$. Values of $<r_x^2>$ required by this equation were computed using a simplified rotational isomeric state model neglecting attractive forces between non-bonded atoms and groups (83, 87).

This illustrative study of ring-chain equilibria in a paraffin-siloxane system demonstrates that the conformational characteristics of chain molecules can be investigated provided suitable labile groups are incorporated into the molecular structures. It should prove within the capabilities of chemists to synthesise a wide variety of molecules suitable for study by the equilibrium cyclic concentration method using the dimethylsiloxane (or other siloxane) linkage as a labile linkage, incorporated into the molecular structure, through which to effect ring-chain equilibration reactions.

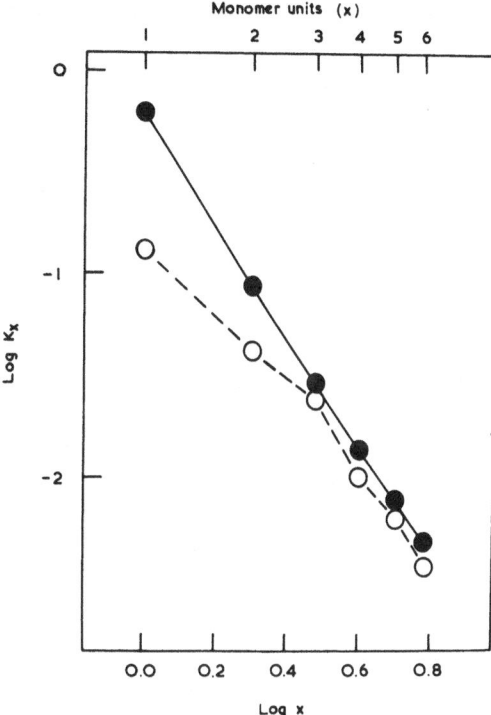

Fig. 9. Experimental molar cyclization equilibrium constants K_x (in mol dm^{-3}) for cyclics $((CH_3)_2Si-(CH_2)_4-(CH_3)_2Si-O)_x$ in an undiluted equilibrate at 298 K (o) are compared with theoretical values calculated by the Jacobson and Stockmayer theory (•) (see text)

C. Block-Copolymers of Polystyrene-Poly(dimethylsiloxane) (88)

The fact that block copolymers of polystyrene and poly(dimethylsiloxane) had already been characterised (89–91), encouraged a preliminary study (88) of the feasibility of studying the statistical conformations of polystyrene chains by using labile dimethylsiloxane linkages to effect ring-chain equilibration reactions. Block copolymers with the structures poly(dimethylsiloxane) poly-b-styrene-b-poly(dimethylsiloxane) were prepared from styrene and hexamethylcyclotrisiloxane by the methods of Wenger (92) and Morton and his co-workers (93). Potassium naphthalene was used as the catalyst for the styrene polymerisation, potassium ions were then exchanged for lithium ions before preparing the terminal dimethylsiloxane blocks. The whole block co-polymer was equilibrated in toluene solution at 383 K by the addition of 0.01% w/v of catalyst (50% w/w potassium hydroxide in suspension in diglyme). The ring-chain equilibration reaction and the analysis of the equilibrate were carried out following the work of Wright (16, 19, 20) on the poly(dimethylsiloxane) system. G.l.c. analysis of the small cyclic siloxanes was used to follow the approach to equilibrium. The equilibrate was quenched by cooling to room temperature and adding small quantities of acetic acid. The ring-chain equilibrate was separated into four fractions using methanol as a precipitating agent. Each fraction was analysed by g.p.c., and proton nuclear magnetic resonance spectroscopy showed that they all contained styrene and dimethylsiloxane linkages.

From his studies, Jones (*88*) concluded that one of the fractions was a mixture of cyclics of average structure

$$(O-(CH_3)_2Si-(styrene)_{25}-(CH_3)_2Si\text{(}O-(CH_3)_2Si)_8)_x.$$

In Fig. 10, a plot of log molarity of the cyclics containing 50, 100, 150 skeletal bonds (x = 1, 2, 3) are plotted against log x and compared with a theoretical line

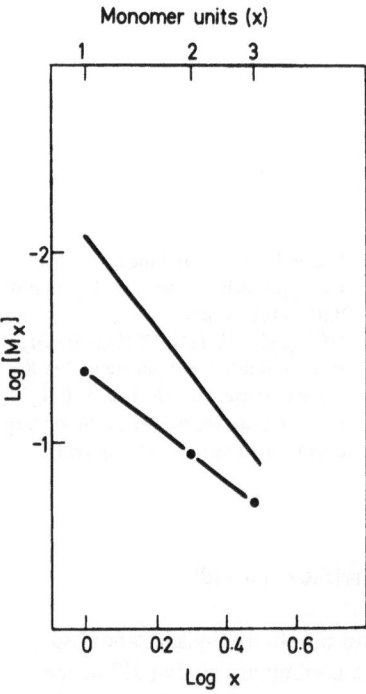

Fig. 10. Concentrations of macrocyclics M_x (in mol dm^{-3}) plotted as log $[M_x]$ against log x for a ring-chain equilibration of poly(dimethylsiloxane)-b-polystyrene-b-poly(dimethylsiloxane) in toluene at 383 K. Experimental points are denoted •. The continuous line was calculated for polystyrene cyclics opening at intervals of fifty skeletal bonds, using the Jacobson and Stockmayer theory and assuming that the chains obey Gaussian statistics

computed by the Jacobson and Stockmayer expression [Eq. (6)] by assuming, for simplicity, that $\sigma_{Rx} = x$ and taking the ratio $<r_x^2>/2xl^2$ for polystyrene to be 10.2[27] and independent of x.

The experimental and theoretical plots shown in Fig. 10 are based on several simplifying assumptions, including some used to interpret g.p.c. tracings in terms of macrocyclic concentrations. Nonetheless, the studies of Jones (*88*) do demonstrate that ring-chain equilibrium reactions can be carried out using suitable block copolymers. However, more work is required to refine the experimental methods. Already, however, attention may be drawn to the large deviations from Gaussian statistics [as embodied in Eq. (5)] shown by polystyrene chains with 50–150 skeletal bonds for the probabilities of their intramolecular cyclization, compared with dimethylsiloxane chains of comparable length (see Fig. 5). Finally, it is noted that studies similar to those of Jones (*88*) for the polystyrene-poly(dimethylsiloxane) system may well prove possible for other block copolymer systems [see, for example, Refs. (*94, 95*)].

D. Polyethers

1. Cyclics in Poly(1,3-dioxolane) (96)

The five membered cyclic 1,3-dioxolane ($CH_2OCH_2CH_2O$) can be polymerised by a variety of catalysts including sulphuric acid (97), perchloric acid (98), phosphorus pentachloride (99) and alkyl aluminium compounds with water as a co-catalyst (100). The effect of the catalyst boron trifluoride diethyl etherate on the polymer-isation of 1,3-dioxolane has also been studied and it has been found that equilibrium between monomeric 1,3-dioxolane and poly(1,3-dioxolane) is set up in both the un-diluted polymer and in solution (101–104). Controversy has arisen as to whether the equilibrium is between cyclic monomer and cyclic polymer (98) or between cyclic monomer and chain polymer (104).

A study of the molecular species in monomer-polymer equilibrates of poly(1,3-dioxolane) was undertaken by Andrews (96, 105). Bulk and solution polymerisa-tions were carried out by adding boron trifluoride diethyl etherate to 1,3-dioxolane alone or to 1,3-dioxolane dissolved in dichloroethane. Equilibria were attained at 333 K and the equilibrates were quenched in diethylamine. Oligomers were extracted from the polymeric products and analysed by g.l.c. The identities of the cyclic dimer and cyclic trimer ($CH_2OCH_2CH_2O)_x$ ($x = 2,3$) were established by mass spectrom-etry. The nuclear magnetic resonance and infra-red spectra of the oligomeric frac-tions corresponded to those expected for the cyclics ($CH_2OCH_2CH_2O)_x$. The re-latively high molecular weight polymer was analysed by g.p.c. and the whole system was found to be a ring-chain equilibrate.

Experimental molar cyclization equilibrium constants for cyclics in the undiluted equilibrate (containing 81% linear polymer) and the solution equilibrate (containing 14% linear polymer) are plotted as log K_x versus log x in Fig. 11. The K_x values are believed to be correct to within ± 10% for cyclics with $x = 2-5$, and it is concluded

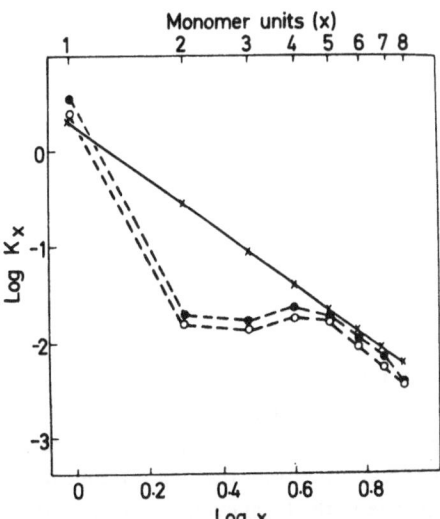

Fig. 11. Experimental molar cyclization equilibrium constants K_x (in mol dm^{-3}) in undiluted (o) and solution (•) equilibrates of poly(1,3-dioxolane) at 333 K compared with values calculated (x) by the Jacobson and Stockmayer theory

that chains adopt random-coil conformations with similar contours in both the un-diluted and the solution equilibrate.

The theoretical K_x values plotted in Fig. 11 were calculated by Eq. (6) with $\sigma_{Rx} = 2x$. Values of $<r_x^2>$ were computed using a rotational isomeric state model based on the models for polyethers by Flory and Mark (106, 107) and giving a characteristic ratio $(<r^2>_0/5xl^2)_\infty$ of 4.4. This is close to Gorin and Monnerie's (108) experimental values of 3.9 at 318 K and 4.2 at 280 K. Finally, it is noted that there is close agreement between experiment and theory for cyclics with 25 or more skeletal bonds (see Fig. 11) showing that 1,3-dioxolane chains of these lengths obey the Gaussian relationship for the probability of intramolecular cyclization.

2. Oligomers in Poly(tetrahydrofurane) (109)

Just as in the poly(1,3-dioxolane) system, monomer-polymer equilibria have been set up between tetrahydrofuran ($CH_2CH_2CH_2CH_2O$) and poly(tetrahydrofuran) both in the undiluted polymer (110–113) and in solution (114, 115). However, by contrast with the poly(1,3-dioxolane) system, poly(tetrahydrofuran) monomer-polymer equilibrates have been found to contain no cyclic oligomers (105, 109). Analysis of oligomeric extracts by g.l.c. showed two series of oligomers, one being in considerably higher concentrations than the other. Mass spectrometry showed that the oligomers from either series were not the cyclics $(CH_2CH_2CH_2CH_2O)_x$. This is in agreement with the fact that the weight fractions of the higher oligomers are considerably lower than those predicted by the Jacobson and Stockmayer theory (21) and are approximately independent of dilution.

The fact that cyclic oligomers are not formed in the poly(tetrahydrofuran) system is surprising in view of the prediction by Dreyfuss and Dreyfuss (116) that macrocyclic oxonium ions would be expected to be formed in such systems; their conclusions being based on kinetic studies (117–119). In this connection it should prove interesting to analyse the oligomeric content of other polyethers. The presence or absence of cyclic oligomers, and their distributions, should prove a useful probe of the mechanisms of the reactions involved.

E. Polyesters

1. Aliphatic Polyesters (120)

Despite the numerous polyesters described in the literature, it is only recently that the cyclic populations of ring-chain equilibrates of linear aliphatic polyesters have been characterised. This characterisation was based on the well-known investigations of Carothers (121, 122) and his co-workers and the studies of Billmeyer (123–126) and his group. In particular, Billmeyer's work established tetraisopropyltitanate as a catalyst for producing equilibrium between ring and chain molecules in polyester systems. This is in contrast to the catalyst, p-toluene sulphonic acid, used by Jacobson, Beckmann and Stockmayer (127) in what was effectively the first attempt to study experimentally and theoretically the change in the total cyclic population of a polymer with solvent dilution. These authors used viscometry to obtain an estimate of the weight fraction of cyclics in ring-chain equilibrates of poly(decamethylene adipate) but no attempt was made to separate individual species.

In a recent study (120), measurements were made of the molar cyclization equilibrium constants K_x for cyclics $(O(CH_2)_{10}OCO(CH_2)_4CO)_x$ with $x = 1-5$ in an undiluted equilibrate of poly(decamethylene adipate) (PDA) at 423 K, and for cyclics $(O(CH_2)_3OCO(CH_2)_2CO)_x$ with $x = 1-7$ in an undiluted equilibrate of poly(trimethylene succinate) (PTS) at the same temperature. The polymers were prepared from dimethyl adipate and decamethylene glycol and from dimethyl succinate and 1,3-propane diol using tetraisopropyltitanate and equilibrated at the required temperature in four-necked glass reaction kettles. Cyclics were extracted from the polymeric equilibrates and analysed by g.p.c. by methods described in Ref. (120). Individual cyclics were also prepared from the polyesters by the general pyrolytic method of Carothers and these were used for identification and calibration purposes.

Experimental molar cyclization equilibrium constants K_x for cyclics in the PDA melt and the PTS melt at 423 K are shown plotted as $\log K_x$ against $\log x$ in Fig. 12 and 13. They are compared with theoretical values calculated by the Jacobson and Stockmayer expression Eq. (6) with $\sigma_{Rx} = 2x$. Values of $<r_x^2>_0$ required by this expression were computed by the exact mathematical methods of Flory and Jernigan (37, 38) using the rotational isomeric state models for the polyesters set up by Flory and Williams (27, 128). Agreement between experiment and theory is excel-

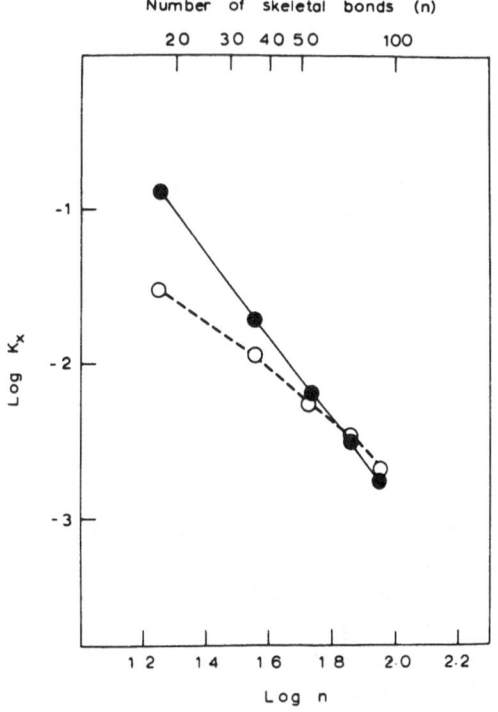

Fig. 12. Experimental molar cyclization equilibrium constants K_x (in mol dm^{-3}) for cyclics $(O(CH_2)_{10}OCO(CH_2)_4CO)_x$ in the PDA melt at 423 K (o) are plotted as $\log K_x$ against $\log n$, where n represents the number of skeletal bonds in the ring. They are compared with theoretical values calculated by the Jacobson and Stockmayer theory (•)

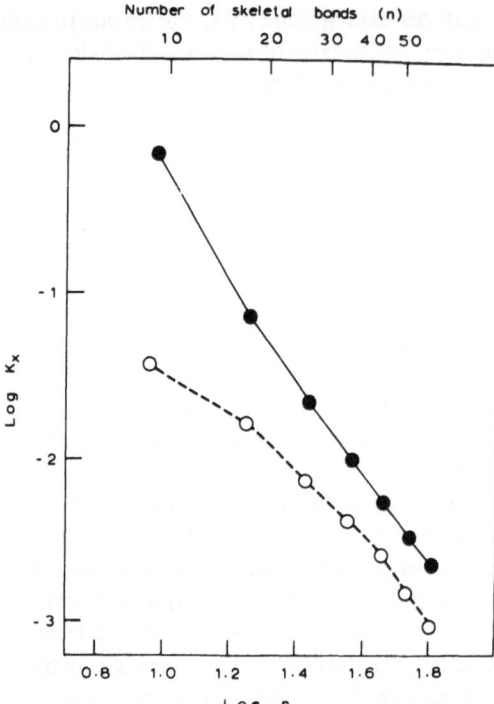

Fig. 13. Experimental molar cycliza-tion equilibrium constants K_x (in mol dm^{-3}) for cyclics $(O(CH_2)_3OCO(CH_2)_2CO)_x$ in the PTS melt at 423 K (o) are plotted as $\log K_x$ against $\log n$, where n represents the number of skeletal bonds in the ring. They are compared with theoretical values calculated by the Jacobson and Stockmayer theory (•)

lent for the largest cyclics analysed in the PDA equilibrate, and K_3, K_4 and K_5 for the cyclics containing 54, 72 and 90 skeletal bonds respectively are within experimental error of the corresponding theoretical K_x values. By contrast, the experimental K_4, K_5, K_6 and K_7 values for the cyclics containing 36, 45, 54 and 63 skeletal bonds respectively in the PTS equilibrate are about half the corresponding theoretical values. Cyclics in the latter system contain twice as many ester groups as cyclics of similar size in PDA, and it is concluded that the conformational characteristics of oligomeric aliphatic polyester chains are considerably more sensitive to the relative numbers of methylene groups and ester linkages than is predicted by the Flory and Williams models (27, 128).

2. Poly(ethylene terephthalate) (129, 130)

Poly(ethylene terephthalate) is commercially the most important of all the polyesters containing aromatic groups. One of its trade names is 'Terylene' and 'Dacron' is another. The polymer consists of linear chains containing the repeat unit $CO \cdot C_6H_4 \cdot CO \cdot O \cdot CH_2 \cdot CH_2 \cdot O$ and it has been found to undergo ring-chain equilibration reactions in the melt and in solution, in the presence of suitable catalysts. Such equilibration reactions have been the subject of recent investigations (17, 130). These investigations were based on several studies, including the extraction of cyclic oligomers from commercial PET samples, and the preparation and characterisation of individual cyclic oligomers. The cyclic trimer $(CO \cdot C_6H_4 \cdot CO \cdot O \cdot CH_2 \cdot CH_2 \cdot O)_3$

is the cyclic present in the highest concentration in commercial PET film (*131, 132*), fibre and chip (*133*); although a cyclic tetramer and a cyclic pentamer have also been detected (*133*). Chromatographic methods of separation have been applied to the cyclics $(CO \cdot C_6H_4 \cdot CO \cdot O \cdot CH_2 \cdot CH_2 \cdot O)_x$ with $x = 3-6$. Individual cyclic oligomers of PET from the dimer to the hexamer have been synthesised (*136–140*) and then characterised using a range of physico-chemical techniques (*139–148*). All these studies provided a basis for the investigation of the cyclic populations of ring-chain equilibrates of PET to be outlined here.

The equilibration reactions were carried out using PET chip of high molecular weight and low cyclic content (*17, 130*). Melt equilibration reactions were carried out at 543 K using antimony trioxide as catalyst. Solution equilibrates in 1-methyl naphthalene were carried out using zinc acetate as catalyst. Cyclic oligomers were extracted from the quenched equilibrates and analysed by g.p.c. Fig. 14 shows a

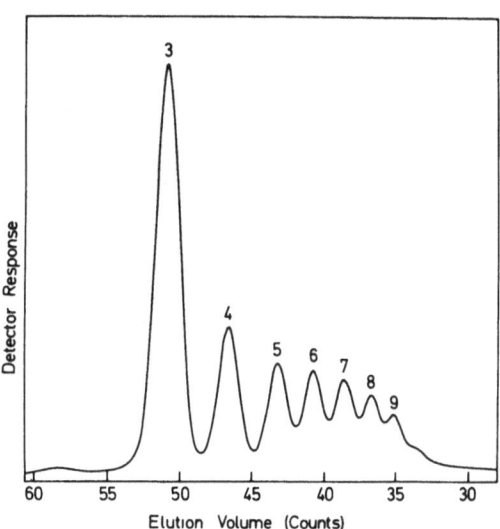

Fig. 14. Gel permeation chromato-gram (g.p.c.) of cyclics $(CO \cdot C_6H_4 \cdot CO \cdot O \cdot CH_2 \cdot CH_2 \cdot O)_x$ with $x = 3-9$ obtained by extracting melt polymerised samples of poly-(ethylene terephthalate) with 1-me-thyl naphthalene

typical chromatogram of a mixture of cyclics $(CO \cdot C_6H_4 \cdot CO \cdot O \cdot CH_2 \cdot CH_2 \cdot O)_x$ from the trimer to the nonamer obtained by extracting a melt polymerised sample.

Experimental molar cyclization equilibrium constants K_x for cyclics in molten PET at 543 K are shown in Fig. 15. They are compared with theoretical values calculated by the Jacobson and Stockmayer theory (*21*), using Eq. (6) with $\sigma_{Rx} = 2x$. Values of $<r_x^2>_0$ required by Eq. (6) were computed by the mathematical methods of Flory and Jernigan (*37, 38*) using the rotational isomeric state model of Williams and Flory (*149*). There are striking discrepancies between the experimental and theoretical values. These discrepancies could arise from a failure of the corresponding open chain molecules to obey Gaussian statistics [see Section II.(B)]; or they could arise from correlations between the positions and directions of the chain termini in their highly-coiled conformations resulting in asymmetric distributions of end-

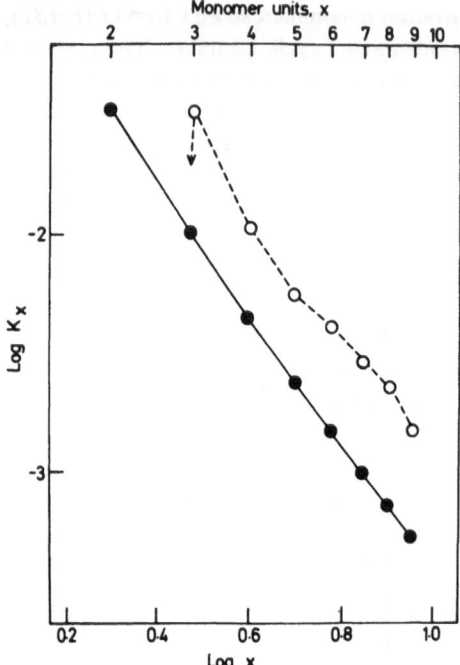

Fig. 15. Experimental molar cyclization constants K_x (in mol dm^{-3}) for cyclics $(CO \cdot C_6H_4 \cdot CO \cdot O \cdot CH_2 \cdot CH_2 \cdot O)_x$ with $x = 2–9$ in the melt at 543 K (denoted ○) are plotted as $\log K_x$ against $\log x$. They are compared with theoretical values calculated by the Jacobson and Stockmayer theory assuming that the corresponding open chain molecules obey Gaussian statistics (these values are denoted ●)

to-end vectors and non-random relative orientations of terminal bonds in the intramolecular cyclization reactions [see Section II.(C)]. Notwithstanding these discrepancies, the results of the investigations show that there is close agreement between the molar concentrations of cyclics $(CO \cdot C_6H_4 \cdot CO \cdot O \cdot CH_2 \cdot CH_2 \cdot O)_x$ with $x = 3–9$ in a melt equilibrate containing 95% w/w linear polymer and in a solution equilibrate containing only 6% w/w linear polymer. Hence PET chains must adopt similar average conformations in the bulk and in dilute solution in 1-methyl naphthalene.

Finally, it is noted that some studies of ring-chain equilibration reactions in solid PET samples have been carried out (*17, 130*). Heating PET chip containing about 2.2% w/w cyclics up to the nonamer resulted in a decrease in cyclic content to about 0.7% w/w, and an increase in molecular weight by a factor of about two. Similar decreases in the concentrations of cyclic oligomers have been found to result when melt equilibrated samples of nylon. 6 are heated in the solid state, and these will be discussed in detail in the next section.

F. Polyamides

1. Cyclic Content of Molten Nylon-6 *(150–155)*

Most of the quantitative studies of the cyclic oligomer concentrations of polyamides have been concerned with nylon-6 (*150–155*), although some preliminary work has been directed at determining the cyclic content of nylon-6,6 (*105, 156–158*). In a recent study (*155*) molar cyclization equilibrium constants K_x for cyclic oligomers $(NH(CH_2)_5CO)_x$ with $x = 1–6$ were measured for an undiluted melt equilibrate of nylon-6 at 525 K. This study was based on the work of a number of authors, who developed methods for preparing equilibrated polymers (*159–162*) and for separating and identifying cyclic oligomers (*163–166*).

The ring-chain equilibrates of nylon-6 at 525 K were prepared starting from ε-caprolactam, and using 6-aminohexanoic acid as a catalyst to give an effective water concentration of 0.2% w/w. The equilibrates were quenched by cooling the melts to room temperature and the polymeric products were turned to small, fine shavings on a lathe. Cyclic oligomers were extracted using hot methanol and analysed by g.l.c. (monomer, dimer and trimer) and by g.p.c. (cyclic oligomers up to the hexamer). Fig. 16 shows a g.p.c. of cyclics $(NH(CH_2)_5CO)_x$ with $x = 1–6$ obtained from

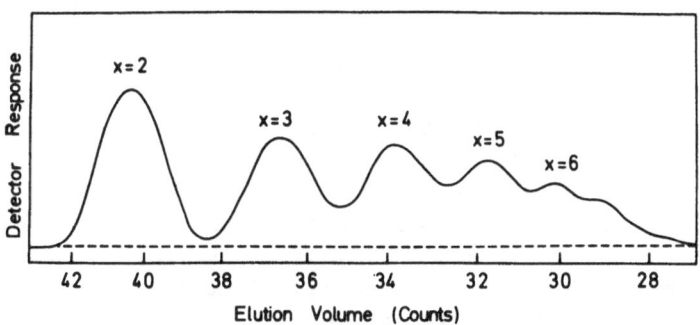

Fig. 16. Gel permeation chromatogram (g.p.c.) of cyclics $(NH(CH_2)_5CO)_x$ with $x = 2–6$ obtained from nylon-6 prepared by ring-chain equilibration in the melt at 525 K

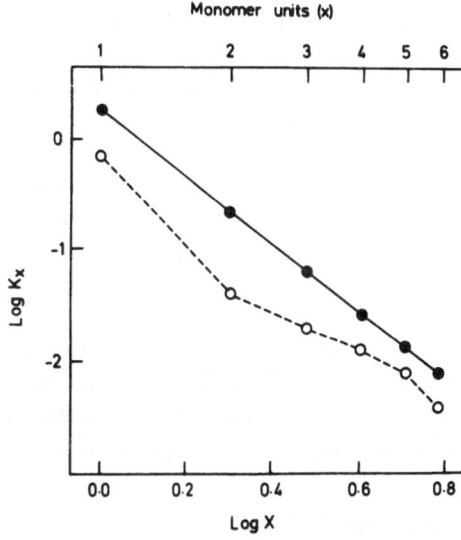

Fig. 17. Experimental molar cyclization equilibrium constants K_x (in mol dm^{-3}) for cyclics M_x in the melt at 525 K (o) are plotted as log K_x against log x. They are compared with theoretical values calculated by the Jacobson and Stockmayer theory which assumes that the corresponding open chain molecules obey Gaussian statistics (these values are denoted ●)

a melt ring-chain equilibrate of nylon-6. In Fig. 17, the experimental molar cyclization equilibrium constants for cyclics $(NH(CH_2)_5CO)_x$ with $x = 1-6$ in molten nylon-6 at 525 K are compared with theoretical values calculated by the Jacobson and Stockmayer theory (21), using Eq. (6) with $\sigma_{Rx} = x$. Values of $<r_x^2>_0$ required by Eq. (6) were computed by the mathematical methods of Flory and Jernigan $(37, 38)$ using the rotational isomeric state model of Flory and Williams (128). The uncertainties in the experimental K_x values for the cyclics from monomer to pentamer are believed to be less than $\pm 10\%$ and so the differences between the experimental and the corresponding theoretical values for K_4, K_5 and K_6 are significant. The experimental values are lower than the corresponding theoretical ones by factors of 2,1.7 and 2 respectively. Similar discrepancies have been found for a polar aliphatic polyester system, but not for a non-polar aliphatic polyester [see Section III.(E)1]

2. Cyclic Concentrations in Samples of Solid Nylon-6 (155)

Wichterle and his co-workers $(167-169)$ established that ring-chain equilibration reactions can be carried out in nylon-6 at temperatures well below the melting point of the polymer[1] [see Section III.(E)2 for similar reactions in poly(ethylene terephthalate)]. Such ring-chain equilibration reactions have been carried out recently (155) by heating samples of nylon-6 or ϵ-caprolactam at 459 K in three clearly defined experiments. The individual cyclic oligomer concentrations were determined for each solid sample of nylon-6. The results obtained are plotted in Fig. 18 and 19, where they are compared with cyclic concentrations in a polymer prepared by ring-chain equilibration in the melt at 525 K (see Fig. 17).

The results shown in Fig. 18 and 19 show conclusively that thermodynamic equilibria between monomeric ϵ-caprolactam and polymeric nylon-6 can be set up in the solid polymer well below its melting point. Similar monomer concentrations are obtained by heating at 459 K, for over 30 days, starting with (i) polymer containing 8.2% w/w monomer, (ii) polymer containing no monomer and (iii) monomer alone.

Fig. 19 shows that the concentrations of cyclic oligomers (with $x = 3-6$) in the nylon-6 prepared by starting with monomer alone (iii), are lower by factors of about four than the corresponding values for melt equilibrated polymer. Following Wichterle $(167-169)$ these differences can be interpreted in terms of a polymer structure at 459 K consisting of about 80% crystalline regions (free from cyclic oligomers) together with about 20% amorphous regions- the latter containing rings in thermodynamic equilibrium with chains, which adopt similar random-coil conformations to those in the molten polymer.

A similar interpretation cannot be put on the cyclic oligomer concentrations of samples prepared starting with solid polymers containing (i) 11.5% w/w cyclic oligomers and (ii) no cyclic oligomers (see Fig. 18). Evidently the morphological structure of the samples of nylon-6 at 459 K prepared by starting with (i) and (ii) are similar,

---

[1] The melting point of nylon-6 is quoted as 488 K by W. Sbrolli (224).

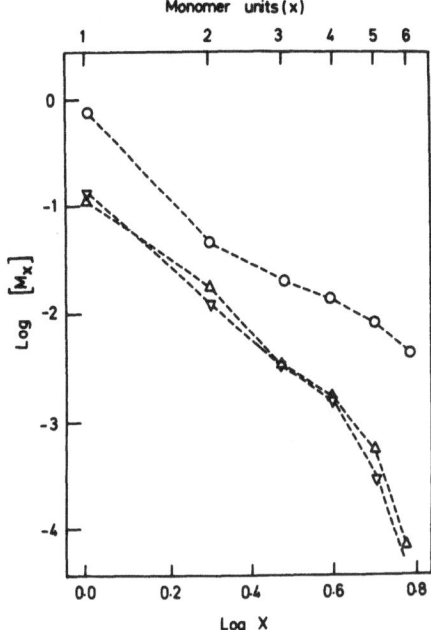

Fig. 18. Concentrations of cyclics M_x (in mol kg^{-1}) in samples of nylon-6 prepared (i) by heating a melt polymerised sample in the solid state at 459 K for 32 days ($\triangle$) and (ii) by heating a cyclic-free polymer in the solid state at 459 K for 34 days (∇). These values are compared with cyclic concentrations in a polymer prepared by ring-chain equilibration in the melt at 525 K ($\circ$)

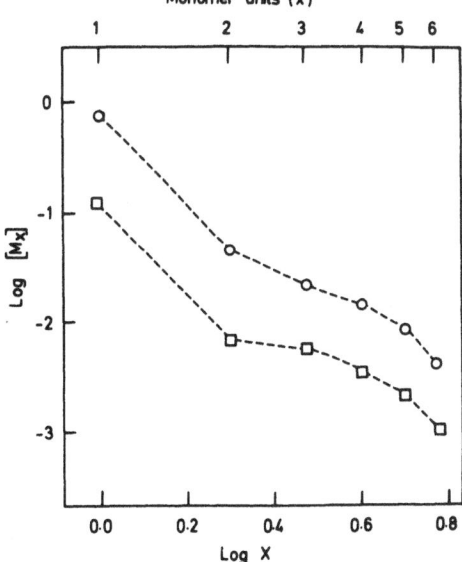

Fig. 19. Concentrations of cyclics M_x (in mol kg^{-1}) in a sample of nylon-6 prepared by polymerising ϵ-caprolactam at 459 K and heating for a total of 36 days ($\square$). These values are compared with cyclic concentrations in a polymer prepared by ring-chain equilibration in the melt at 525 K ($\circ$)

and both are very different from that prepared by heating (iii) the cyclic monomer ϵ-caprolactam. This study suggests that the determination of the individual cyclic concentrations in polymers prepared by ring-chain equilibration reactions in the solid state provides a sensitive probe of the structure and morphology of the polymer [in this connection see also Refs. (*170–173*)].

G. Polyphosphates

1. Cyclics in Sodium Phosphate Melts (174–176)

Graham's salt (or sodium phosphate) is a glass with the empirical formula $NaPO_3$ (177). It may be prepared by heating sodium dihydrogen phosphate at high temperatures for many hours and then quenching the melt rapidly between metal plates (178, 179). The quenched product is soluble in water. It has been extensively investigated (180–183) and shown to consist of long linear polyphosphate chains, terminated by hydroxyl groups (184, 185), together with about 10% w/w cyclics $(NaPO_3)_x$ with $x = 3$–7 (174, 175). A ring-chain equilibrium is established in the melt and the concentrations of cyclics in quenched equilibrates have been determined by van Wazer and his co-workers (for $x = 3$–6) (174, 186) and by Thilo and Schülke (for $x = 3$–7) (175). Both groups used paper chromatography as their main analytical technique.

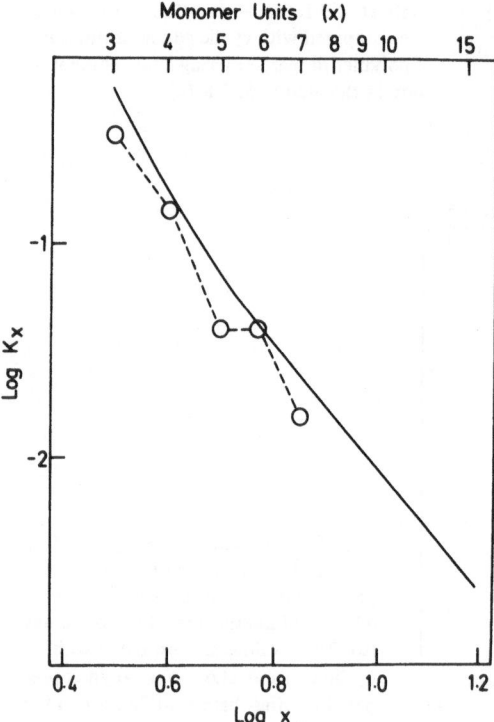

Fig. 20. Experimental molar cyclization equilibrium constants K_x (in mol dm^{-3}) for oligomeric sodium metaphosphates $(NaPO_3)_x$ in melts at 1000 K (denoted o) are compared with theoretical values calculated by the Jacobson and Stockmayer theory assuming that all the chains obey Gaussian statistics (denoted by the unbroken line)

In Fig. 20, experimental molar cyclization equilibrium constants K_x for sodium metaphosphates $(NaPO_3)_x$ in molten sodium phosphate at 1000 K are compared with theoretical values (176). The latter were calculated by the Jacobson-Stockmayer theory (21) assuming that chains of all lengths obey Gaussian statistics and using

Eq. (6) with $\sigma_{Rx} = 2x$. A rotational isomeric state model for polyphosphate chains (*176, 187*) was used to compute values of $< r_x^2 >_0$ required by Eq. (6). Comparison of the experimental values with the theoretical values, as well as with the K_x values for the corresponding cyclic siloxanes $(R(CH_3)SiO)_x$ [see Section III.(A) 5] strongly supports the contention that polyphosphate chains in sodium phosphate melts adopt random-coil conformations. Further evidence for the structure of sodium phosphate melts would come from precise data for macrocyclic concentrations in quenched sodium phosphate ring-chain equilibrates, if these became available.

2. Other Phosphate Systems

In connection with the outline of experimental and theoretical work on sodium phosphate melts, it is appropriate to mention some studies on other phosphate systems. For example, van Wazer and his co-workers (*188*) have recently published ^{31}P nuclear magnetic resonance investigations of ring-chain equilibria of some phosphoric acids in organic solvents. These investigations are related to a number of theoretical and experimental studies of molecular equilibria in inorganic systems that have been carried out by van Wazer and his co-workers (*189–196*). Although many of these studies are concerned with non-linear systems, some yield information relating to cyclic concentrations. However, none of them give precise molar cyclization equilibrium constants for individual rings, which could be related directly to the conformations of the corresponding open chain molecules. As an example of this, an investigation of ring-chain equilibria in the system $POCl_3/P_2O_5$ may be cited. This was studied experimentally by van Wazer and his co-workers (*197, 198*) and it has received an extensive theoretical treatment by Gordon and Scantlebury (*9*). The polycondensation process was developed in detail in terms of cascade theory (or the theory of branching processes) (*6, 199–202*) and only a limited discussion of cyclic formation in the system was made. For all such studies of ring formation in non-linear systems see Ref. (*1*).

H. Polymeric Sulphur

The presence of cyclic molecules in liquid sulphur has been the subject of speculation for many years. It is now well established that liquid sulphur contains substantial amounts of cyclooctasulphur S_8 (*203*); and recently quantities of cyclododecasulphur S_{12} have been recovered from liquid sulphur by Schmidt and Block (*204*). It is widely recognised that there is a ring-chain equilibrium in the liquid element above a critical transition temperature of 432 K, which involves cyclooctasulphur and diradical chains, some with a very high average degree of polymerisation. This equilibrium has been the subject of theoretical studies by Gee (*205*) and by Tobolsky and Eisenberg (*206*).

In this review, attention will be directed to the possibility of a cyclic population in liquid sulphur comparable to the other ring-chain equilibria that have been discussed previously. In particular, consideration will be given to the constitution of liquid sulphur between the melting-point of the solid element and the critical transition temperature. In this region, elemental sulphur exhibits a self-depression of its freezing point, which was discovered by Gernez (*207*) a hundred years ago. His discovery has prompted experimental investigations (*208–211*) and theoretical interpretations of the phenomenon (*212–214*). The most comprehensive experimental investigation has been by Wiewiorowski and his co-workers (*215*). They found that liquid sulphur froze at temperatures as low as 7.6 K below the melting point of monoclinic sulphur. There has been much speculation as to the identity of the molecular species causing this large freezing point depression. Since Aten (*212*) called the species S_π and identified it as a cyclic S_4, there have been many other suggestions as to its nature. These include postulations that S_π is mainly the cyclic S_6 (*216*) that it consists of diradical chains containing eight sulphur atoms (*217, 218*), that it is a mixture of small as well as large sulphur cyclics (*213, 219*) and that it consists almost entirely of large cyclic molecules starting from S_9 (*214*). Quantitative theoretical approaches as to the nature of S_π have been made by the present author (*214*) [see also Refs. (*220–222*)] and by Harris (*213*). However a full explanation of the phenomenon must await detailed experimental analysis of the liquid element.

I. Conclusion

This review has been concerned with the concentrations of ring molecules in ring-chain equilibrates of a wide range of polymers. Emphasis has been placed on systems where molar cyclization equilibrium constants K_x for large ring molecules are known. Only in the case of polar polymers (such as nylon-6) and those containing polystyrene are there appreciable differences between known K_x values and those calculated by the simple Jacobson and Stockmayer theory in the range of large cyclics. In general, the Gaussian assumption of the Jacobson and Stockmayer theory would not be expected to be appropriate for small or medium sized molecules and new theoretical approaches must be made [see Refs. (*87, 223*)]. Such approaches will not be discussed here. The primary aim of this review is to show how the study of equilibrium cyclic concentrations provides detailed structural information for a wide range of molecules. Furthermore, such studies can be used as an experimental method for probing the average conformations of chain molecules in a variety of different environments including liquid and solid polymers and polymer solutions.

Acknowledgements. It is a pleasure to take the opportunity of thanking some of those who were involved in the research carried out with the author at the University of York, part of which has been discussed here in relation to published contributions of other authors. Much of the work at York was carried out by Dr. P. V. Wright, Mr. G. R. Walker, Dr. M. S. Beevers, Dr. J. M. Andrews, Dr. D. R. Cooper, Dr. F. R. Jones, Dr. L. E. Scales and Mr. D. Sympson. Assistance by advice and by the gift of materials was generously provided by research groups at the Dow Corning Corporation, Barry, Wales, at Courtaulds Ltd., London, England and at I. C. I. Fibres Ltd., Harrogate, England. Finally sincere thanks are extended to Professor P. J. Flory of Stanford University, California, USA, who was a most inspiring teacher for the author at the beginning of his involvement with studies of cyclics in polymers, and who has provided valuable ideas and advice ever since.

IV. References

1. Stepto, R. F. T.: Adv. Polym. Sci.: to be published.
2. Kilb, R. W.: J. Phys. Chem. **62**, 969 (1958).
3. Frisch, H. L.: paper presented at 128th meeting Amer. Chem. Soc., Minneapolis 1955.
4. Stepto, R. F. T.: Discussions of the Faraday Society, Gels and Gelling Processes, Norwich 1974.
5. Gordon, M., and Scantlebury, G. R.: J. Chem. Soc. **B1** (1967).
6. Gordon, M.: Proc. Roy. Soc. **A272**, 240 (1962).
7. Peniche-Covas, C. A. L., Dev, S. B., Gordon, M., Judd, M., and Kajiwara, K.: Discussions of the Faraday Society, Gels and Gelling Processes, Norwich 1974.
8. Gordon, M., and Temple, W. B.: Makromol. Chem. **160**, 263 (1972).
9. Gordon, M., and Scantlebury, G. R.: Proc. Roy. Soc. **A292**, 380 (1966).
10. Hopkins, W., Peters, R. H., and R. F. T. Stepto: Polymer **15**, 315 (1974).
11. Peters, R. H., and Stepto, R. F. T. in: The chemistry of polymerisation processes. Monograph No. 20, Soc. Chem. Ind., London 1965.
12. Stepto, R. F. T., and Waywell, D. R.: Makromol. Chem. **152**, 263 (1972).
13. Gaetjens, E., and Morawetz, H.: J. Amer. Chem. Soc. **83**, 1738 (1961).
14. Goodman, N, and Morawetz, H.: J. Polym. Sci. (C) **31**, 177 (1970).
15. Sisido, M.: Polym. J. **3, 1**, 84 (1972).
16. Wright, P. V., Phil, D.: thesis, University of York (1970).
17. Cooper, D. R.: D. Phil. thesis, University of York (1972).
18. Brown, J. F., and Slusarczuk, G. M.,J.: J. Amer. Chem. Soc. **87**, 931 (1965).
19. Semlyen, J. A., and Wright, P. V.: Polymer **10**, 543 (1969).
20. Wright, P. V.: J. Polym. Sci., Polym. Physics Edition **11**, 51 (1973).
21. Jacobson, H., and Stockmayer, W. H.: J. Chem. Phys. **18**, 1600 (1950).
22. Tanford, C.: Physical chemistry of macromolecules. New York: J. Wiley, 1961.
23. Morawetz, H.: Macromolecules in solution. New York: Interscience 1965.
24. Tsvetkov, V. N., Eskin, V. E., and Frenkel, S. Ya: Structure of macromolecules in solution, Vol. 1–3. National Lending Library 1971.
25. Cowie, J. M. G.: Polymers: chemistry and physics of modern materials. London: Intertext Books 1973.
26. Flory, P. J.: Principles of polymer chemistry. Ithaca: Cornell University Press 1953.
27. Flory, P. J.: Statistical mechanics of chain molecules. New York: Interscience 1969.
28. Mark, J. E.: Rubber Chem. and Tech. **46**, 593 (1973).
29. Kirste, R. G., Kruse, W. A., and Schelten, J.: J. Makromol. Chem. **162**, 299 (1973).
30. Benoit, H., Cotton, J. P., Decker, D., Farnoux, B., Higgins, J. S., Jannink, G., Ober, R., and Picot, C.: Nature (London) **245**, 13 (1973).
31. Ballard, D. G. H., Schelten, J. and Wignall, G. D.: Eur. Polym. J. **9**, 965 (1973).
32. Ober, R., Cotton, J. P., Farnoux, B., and Higgins, J. S.: Macromolecules **7**, 634 (1974).
33. Lieser, G., Fischer, E. W., and Ibel, K.: Polymer Letters **13**, 39 (1975).
34. Krigbaum, W. R., and Godwin, R. W.: J. Chem. Phys. **43**, 4523 (1965).
35. Brady, G. W., Wasserman, E., and Wellendorf, J.: J. Chem. Phys. **47**, 855 (1967).
36. Flory, P. J., and Semlyen, J. A.: J. Amer. Chem. Soc. **88**, 3209 (1966).
37. Flory, P. J.: Proc. Nat. Acad. Sci., Wash. **51**, 1060 (1964).
38. Flory, P. J., and Jernigan, R. L.: J. Chem. Phys. **42**, 3509 (1965).
39. Volkenstein, M. V.: Configurational statistics of polymeric chains. New York: Interscience 1963.
40. Birshtein, T. M., and Ptitsyn, O. B.: Conformations of macromolecules. New York: Intersience 1966.
41. Lifson, S.: J. Chem. Phys. **30**, 964 (1959).
42. Nagai, K.: J. Chem. Phys. **31**, 1169 (1959).
43. Hoeve, C. A. J.: J. Chem. Phys. **32**, 888 (1960).
44. Flory, P. J.: Macromolecules **7**, 381 (1974).
45. Kuhn, W.: Kolloid-Z. **68**, 2 (1934).

46. Treloar, L. R. G.: The physics of rubber elasticity. Oxford: Clarendon Press 1958.
47. Wright, P. V., and Semlyen, J. A.: Polymer 11, 462 (1970).
48. Beevers, M. S., and Semlyen, J. A.: Polymer 12, 373 (1971).
49. Scott, D. W.: J. Amer. Chem. Soc. 68, 2294 (1946).
50. Hartung, H. A., and Camiolo, S. M.: Division of Polymer Chemistry, 141st National Meeting of the American Chemical Society, Washington, D. C., March 1962.
51. Carmichael, J. B., and Kinsinger, J. B.: Canad. J. Chem. 42, 1996 (1964).
52. Carmichael, J. B., and Winger, R.: J. Polym. Sci. A, 3, 971 (1965).
53. Carmichael, J. B., Gordon, D. J., and Isackson, F. J.: J. Phys. Chem. 71, 2011 (1967).
54. Carmichael, J. B.: J. Macromol. Chem. 1, 207 (1966).
55. Flory, P. J., Crescenzi, V., and Mark, J. E.: J. Amer. Chem. Soc. 86, 146 (1964).
56. Mark, J. E., and Flory, P. J.: J. Amer. Chem. Soc. 86, 138 (1964).
57. Crescenzi, V., and Flory, P. J.: J. Amer. Chem. Soc. 86, 141 (1964).
58. Kargin, V. A., Kitaigorodskii, A. I., and Slonimskii, G. L.: Kolloidn. Zh. 19, 131 (1957).
59. Hoseman, R.: J. Polym. Sci. C, 20, 1 (1967).
60. Pechhold, W.: J. Polym. Sci. C, 32, 123 (1970).
61. Yeh, G. S. Y.: Polymer Preprints, Amer. Chem. Soc., Div. Polymer Chem. 10, 1020 (1969).
62. Flory, P. J.: J. Chem. Phys. 17, 303 (1949).
63. Edwards, S. F.: Proc. Phys. Soc., London 85, 613 (1965).
64. Flory, P. J., and Fisk, S.: J. Chem. Phys. 44, 2243 (1966).
65. Allcock, H. R.: Heteroatom ring systems and polymers. New York: Academic Press 1967.
66. Gee, G. in: Inorganic polymers. Chemical Society publication 1961.
67. Gimblett, F. G. R.: Inorganic polymer chemistry. London: Butterworths 1963.
68. Moedritzer, K.: Adv. Organometal. Chem. 6, 171 (1968).
69. Allen, G., Lewis, C. J., and Todd, S. M.: Polymer 11, 31 (1970).
70. Soulen, J. B., and Silverman, M. S.: J. Polym. Sci. A, 1, 823 (1963).
71. Moedritzer, K.: J. Organometal. Chem. 21, 315 (1970).
72. Lee, C. L.: J. Organometal. Chem. 6, 620 (1966).
73. Davies, W. G.: Midland Silicones Limited, personal communication to P. V. Wright and J. A. Semlyen [see Ref. (47)].
74. Brown, E. D., and Carmichael, J. B.: J. Polym. Sci. B, 3, 473 (1965).
75. Lewis, R. N.: J. Amer. Chem. Soc. 70, 1115 (1948).
76. Daudt, W. H., and Hyde, J. F.: J. Amer. Chem. Soc. 74, 386 (1952).
77. Young, C. W., Servais, P. C., Currie, C. C., and Hunter, M. J.: J. Amer. Chem. Soc. 70, 3758 (1948).
78. Hickton, H. J., Holt, A., and Jarvie, A. W.: J. Chem. Soc. C, 149 (1966).
79. Ivin, K. J. and Leonard, J.: European Polym. J. 6, 331 (1970).
80. Leonard, J.: Macromolecules 2, 661 (1969).
81. Bywater, S.: Trans. Faraday Soc. 51, 1267 (1955).
82. Scott, R. L.: J. Phys. Chem. 69, 261 (1965).
83. Beevers, M. S., and Semlyen, J. A.: Polymer 13, 523 (1972).
84. Piccoli, W. A., Haberland, G. G., and Merker, R. L.: J. Amer. Chem. Soc. 82, 1883 (1960).
85. Kumada, M., and Habuchi, A.: Inst. Polytech. Osaka City University, 3 C, 65 (1952).
86. Sommer, L. H., and Ansul, G. R.: J. Amer. Chem. Soc. 77, 2482 (1955).
87. Beevers, M. S.: D. Phil. thesis, University of York (1972).
88. Jones, F. R.: European Polym. J. 10, 249 (1974).
89. Greber, G. and Metzinger, L.: Makromol. Chem. 39, 167 (1960).
90. Morton, M., Rembaum, A. A., and Bostick, E. E.: J. Appl. Polym. Sci. 8, 2707 (1964).
91. Saam, J. C., Gordon, D. J., and Lindsey, S.: Macromolecules, 3, 1 (1970).
92. Wenger, F.: Makromol. Chem. 64, 151 (1963).
93. Morton, M. Milkovich, R., McIntyre, D. B., and Bradley, L. J.: J. Polym. Sci. A, 1, 443 (1963).
94. Vaughn, H. A.: Polymer Letters 7, 569 (1969).
95. Szwarc, M.: Carbanions, living polymers and electron transfer processes. New York: Interscience 1968.

96. Andrews, J. M., and Semlyen, J. A.: Polymer 13, 141 (1972).
97. Gresham, W. F.: U. S. Patent 2, 394, 910 (1946).
98. Plesch, P. H., and Westerman, P. H.: J. Polym. Sci. C, 3837 (1968).
99. Muetterties, E. L.: U. S. Patent 2, 856, 370 (1958).
100. Okada, M., Yamashita, Y., and Ishii, Y.: Makromol. Chem. 80, 196 (1964).
101. Yamashita, Y., Okada, M., Suyama, K., and Kasahara, H.: Makromol. Chem. 114, 146 (1968).
102. Yamashita, Y., Asakura, T., Okada, M., and Ito, K.: Makromol. Chem. 129, 1 (1969).
103. Gorin, S., and Monnerie, L.: Bull. Soc. Chim. France 2, 2047 (1966).
104. Kelen, T., Schlotterbeck, D., and Jaacks, V.: IUPAC Conference on Macromolecules, Boston, USA 1971.
105. Andrews, J. M.; D. Phil. thesis, University of York (1972).
106. Flory, P. J., and Mark, J. E.: Makromol. Chem. 11, 75 (1964).
107. Mark, J. E., and Flory, P. J.: J. Amer. Chem. Soc. 87, 1415 (1965).
108. Gorin, S., and Monnerie, L.: J. Chim. Phys. 65, 2084 (1968).
109. Andrews, J. M., and Semlyen, J. A.: Polymer 12, 642 (1971).
110. Sims, D.: J. Chem. Soc. 1964, 864.
111. Bawn, C. E. H., Bell, R. M., and Ledwith, A.: Polymer 6, 95 (1965).
112. Rozenberg, B. A., Chekhuta, O. M., Lyudvig, Ye. B., Gantmakher, A. R., and Medvedev, S. S.: Polym. Sci. U.S.S.R. 6, 2246 (1964).
113. Ivin, K. J., and Leonard, J.: Polymer 6, 621 (1965).
114. Vofsi, D., and Tobolsky, A. V.: J. Polym. Sci. A, 3, 2361 (1965).
115. Sims, D.: Makromol. Chem. 98, 235 (1966).
116. Dreyfuss, P., and Dreyfuss, M. P.: Adv. Polym. Sci. 4, 528 (1967).
117. Dreyfuss, M. P., and Dreyfuss, P.: 159th Meeting of Amer. Chem. Soc., Houston, Texas, February 1970.
118. Dreyfuss, P., and Dreyfuss, M. P.: Adv. Chem. Ser. 91, 335 (1969).
119. Dreyfuss, M. P., and Dreyfuss, P.: J. Polym. Sci. A1, 4, 2179 (1966).
120. Jones, F. R., Scales, L. E., and Semlyen, J. A.: Polymer 15, 738 (1974).
121. Hill, J. W., and Carothers, W. H.: J. Amer. Chem. Soc. 55, 5031 (1933).
122. Spanagel, E. W., and Carothers, W. H.: J. Amer. Chem. Soc. 57, 929 (1935).
123. Zavaglia, E. A., Mosher, W. A., and Billmeyer, F. W.: Off. Dig. Fed. Soc. Paint Tech. 33, 229 (1965).
124. Chang, P. S., Zavaglia, E. A., and Billmeyer, F. W.: Off. Dig. Fed. Soc. Paint Tech. 33, 235 (1965).
125. Billmeyer, F. W., and Eckard, A. D.: Macromolecules 2, 103 (1969).
126. Billmeyer, F. W., and Katz, I.: Macromolecules 2, 105 (1969).
127. Jacobson, H., Beckmann, C. O., and Stockmayer, W. H.: J. Chem. Phys. 18, 1607 (1950).
128. Flory, P. J., and Williams, A. D.: J. Polym. Sci. A2, 5, 399 (1967).
129. Walker, G. R., and Semlyen, J. A.: Polymer 11, 472 (1970).
130. Cooper, D. R., and Semlyen, J. A.: Polymer 14, 185 (1973).
131. Ross, S. D., Coburn, E. R., Leach, W. A., and Robertson, W. B.: J. Polym. Sci. 13, 406 (1954).
132. Giuffria, R.: J. Polym. Sci. 49, 427 (1961).
133. Goodman, I., and Nesbitt, B. F.: Polymer 1, 384 (1960).
134. Zahn, H., and Kusch, P.: Z. Gesamte Text.-Ind. 69, 880 (1967).
135. Peebles, L. H., Huffman, M. W., and Ablett, C. T.: J. Polym. Sci. A1,7, 479 (1969).
136. Mersakentis, E., and Zahn, H.: J. Polym. Sci. A1, 4, 1890 (1966).
137. Hamb, L., and Trent, L. C.: J. Polym. Sci. B, 5, 1057 (1967).
138. Mersakentis, E., and Zahn, H.: Chem. Ber. 103, 3034 (1970).
139. Zahn, H., and Repin, J. F.: Chem. Ber. 103, 3041 (1970).
140. Repin, J. H., and Papanikolau, E.: J. Polym. Sci. A1, 7, 3126 (1969).
141. Seidel, B.: Z. Elektrochem. 62, 214 (1958).
142. Grime, D., and Ward, I. M.: Trans. Faraday Soc. 54, 959 (1958).
143. Ward, I. M.: Chem. Ind. 1956, 905; Chem. Ind. 1957, 1102.

144. Binns, G. L., Frost, J. S., Smith, F. S., and Yeadon, E. C.: Polymer **7**, 583 (1966).

145. Ito, E., and Okajima, S.: J. Polym. Sci. **B7**, 483 (1969).

146. Ito, E., and Okajima, S.: Polymer **12**, 650 (1971).

147. Hashimoto, S., and Sakai, J.: Kobunshi Kagaku **23**, 422 (1966).

148. Hashimoto, S., and Jinnai, S.: Kobunshi Kagaku **24**, 36 (1967).

149. Williams, A. D., and Flory, P. J.: J. Polym. Sci. **A2**, 5, 417 (1967).

150. Spoor, H., and Zahn, H.: Z. analyt. Chem. **168**, 190 (1959).

151. Zahn, H., Kunde, J. and Heidemann, G.: Makromol. Chem. **43**, 220 (1961).

152. Rothe, I., and Rothe, M.: Chem. Ber. **88**, 284 (1955).

153. Rothe, M.: J. Polym. Sci. **30**, 227 (1958).

154. Semlyen, J. A., and Walker, G. R.: Polymer **10**, 597 (1969).

155. Andrews, J. M., Jones, F. R., and Semlyen, J. A.: **15**, 420 (1974).

156. Zahn, H.: Z. Gesamte Text.-Ind. **66**, 928 (1964).

157. Mori, S., Furusawa, M., and Takeuchi, T.: Anal. Chem. **42**, 661 (1970).

158. Mori, S., and Takeuchi, T.: J. Chromatog. **49**, 230 (1970).

159. Hermans, P. H.: J. Appl. Chem. **5**, 493 (1955).

160. Reimscheussel, H. K.: J. Polym. Sci. **41**, 457 (1959).

161. Wichterle, O., Sittler, E., Cefelin, P.: Coll. Czech. Chem. Commun. **24**, 2356 (1959).

162. Macchi, E. M., Morosoff, N., and Morawetz, H.: J. Polym. Sci. **A1**, 6, 2033 (1968).

163. van Velden, P. F., van der Want, G. M., Heikens, G. M., Druissink, Ch. A., Hermans, P. H., and Staverman, A. J.: Recueil **74**, 1376 (1955).

164. Hermans, P. H.: Recueil **72**, 798 (1953).

165. Kusch, P., and Zahn, H.: Angew. Chem. **4**, 696 (1965).

166. Mulder, J. L., and Buytenhys, F. A.: J. Chromatog. **51**, 459 (1970).

167. Wichterle, O., Sebenda, J., and Kralicek, J.: Fortschr. Hochpolym. Forsch. **2**, 578 (1961).

168. Wichterle, O.: Makromol. Chem. **35**, 174 (1960).

169. Tomka, J., Sebenda, J., and Wichterle, O.: J. Polym. Sci. C, **16**, 53 (1967).

170. Wunderlich, B., and Liberti, F.: Bull. Am. Phys. Soc. **11**, 248 (1966).

171. Liberti, F., and Wunderlich, B.: J. Polym. Sci. **A2**, 6, 833 (1968).

172. Wunderlich, B.: Macromolecular; Vol 1. New York: Academic Press 1973.

173. Dreyfuss, P., and Keller, A.: J. Macromol. Sci. **B**, 4, 811 (1970).

174. McCullough, J. F., van Wazer, J. R., and Griffith, E. J.: J. Amer. Chem. Soc. **78**, 4528 (1956).

175. Thilo, E., and Schulke, U.: Z. Anorg. Allgem. Chem. **341**, 293 (1965).

176. Cooper, D. R., and Semlyen, J. A.: Polymer **13**, 414 (1972).

177. Graham, T.: Phil. Trans. R. Soc. **A123**, 253 (1833).

178. van Wazer, J. R.: Phosphorus and its compounds. Vol. 1. New York: Interscience 1958.

179. Thilo, E. in Advances in inorganic and radiochemistry (Eds. H. J. Emeleus and A. G. Sharpe), Vol. 4. New York: Academic Press 1962.

180. Lamm, O., and Malmgren, H.: Z. Anorg. Allgem. Chem. **245**, 103 (1940).

181. Samuelson, O.: Sven. Kem. Tidskr. **61**, 76 (1949).

182. van Wazer, J. R.: J. Amer. Chem. Soc. **72**, 639, 644, 647, 655, 906 (1950).

183. Strauss, U. P., Smith, E. H., and Wineman, P. L.: J. Amer. Chem. Soc. **75**, 3935 (1953).

184. Samuelson, O.: Sven. Kem. Tidskr. **56**, 343 (1944).

185. Treadwell, W. D., and Leutwyler, F.: Helv. Chim. Acta. **20**, 931 (1937).

186. van Wazer, J. R. and Karl-Kroupa, E.: J. Amer. Chem. Soc. **78**, 1772 (1956).

187. Semlyen, J. A., and Flory, P. J.: Trans. Faraday Soc. **62**, 2622 (1966).

188. Costello, A. J. R., Glonek, T., Myers, T. C., and van Wazer, J. R. Inorg. Chem. **13**, 5, 1225 (1974).

189. Matula, D. W., Groenweghe, L. C. D., and van Wazer, J. R.: J. Chem. Phys. **41**, 3105 (1964).

190. Matula, D. W., and van Wazer, J. R.: J. Chem. Phys. **46**, 3123 (1967).

191. Levy, R. M., and van Wazer, J. R.: J. Chem. Phys. **45**, 1824 (1966).

192. van Wazer, J. R.: Ann. N. Y. Acad. Sci. **159**, 1, 5 (1969).

193. van Wazer, J. R., and Moedritzer, K.: Angew. Chem., Int. Ed. Engl. **5**, 341 (1966).

194. van Wazer, J. R.: Inorg. Macromol. Rev. **1**, 89 (1970).

195. Moedritzer, K., and van Wazer, J. R.: Inorg. Chem. 4, 1753 (1965).
196. Moedritzer, K., and van Wazer, J. R.: J. Amer. Chem. Soc. 87, 2360 (1965).
197. Groenweghe, L. C. D., Payne, J. H., and van Wazer, J. R.: J. Amer. Chem. Soc. 82, 5305 (1960).
198. Groenweghe, L. C. D., van Wazer, J. R., and Dickinson, A. W.: Analyt. Chem. 36, 303 (1964).
199. Harris, F. E.: The theory of branching processes. Berlin: Springer 1963.
200. Good, I. J.: Proc. Camb. Phil. Soc. 45, 360 (1949).
201. Dobson, G. R., and Gordon, M.: J. Chem. Phys. 41, 2389 (1964).
202. Gordon, M., and Scantlebury, G. R.: Trans. Faraday Soc. 60, 604 (1964).
203. Gee, G.: Sci. Prog. 170, 193 (1955).
204. Schmidt, M., and Block, H. D.: Angew. Chem. 79, 944 (1967).
205. Gee, G.: Trans. Faraday Soc. 48, 515 (1952).
206. Tobolsky, A. V., and Eisenberg, A.: J. Amer. Chem. Soc. 81, 780 (1959).
207. Gernez, M. D.: Compt. rend. 82, 115 (1876).
208. Schaum, K.: Ann. 308, 18 (1899).
209. Smith, A., and Holmes, W. B.: J. Amer. Chem. Soc. 27, 979 (1905).
210. Smith, A., and Carson, C. M.: Z. physik. Chem. 57, 685 (1907).
211. Beckmann, E., Paul, R., and Liesche, O.: Z. anorg. allgem. Chem. 103, 189 (1918).
212. Aten, A. W. H.: Z. physik. Chem. 88, 321 (1914).
213. Harris, R. E.: J. Phys. Chem. 74, 3102 (1970).
214. Semlyen, J. A.: Polymer 12, 383 (1971).
215. Wiewiorowski, T. K., Parthasarathy, A., and Slaten, B. L.: J. Phys. Chem. 72, 1890 (1968).
216. Schenk, P. W., and Thümmler, U.: Z. Electrochem., Ber. Bunsenges. physik. Chem. 63, 1002 (1959).
217. Schenk, P. W., and Thümmler, U.: Z. anorg. allgem. Chem. 315, 271 (1962).
218. Wiewiorowski, T. K., and Touro, F. J.: J. Phys. Chem. 70, 3528 (1966).
219. Krebs, K., and Beine, H.: Z. anorg. allgem. Chem. 355, 113 (1967).
220. Semlyen, J. A.: Trans. Faraday Soc. 63, 743 (1967).
221. Semlyen, J. A.: Trans. Faraday Soc. 63, 2342 (1967).
222. Semlyen, J. A.: Trans. Faraday Soc. 64, 1396 (1968).
223. Scales, L. E.: D. Phil. thesis, University of York (1975).
224. Sbrolli, W. in Man-made Fibres, Science and Technology, Editors H. F. Mark, S. M. Atlas and E. Cernia, Vol. 2, Interscience, New York, 1962.

Received January 27, 1976.

Asymmetric Reactions of Synthetic Polypeptides

Shohei Inoue

Department of Synthetic Chemistry, Faculty of Engineering, University of Tokyo, Bunkyo-ku, Tokyo, Japan

Contents

Introduction

Enzymatic reactions are characterized by their high stereospecificity. The enzyme exhibits activity only towards one of the optical antipodes of the substrate. In reactions where an asymmetric center is newly formed, the product is only one of the optical antipodes. Such a high stereospecificity in the reaction is a reflection of the asymmetric primary and higher order structures of the enzyme molecule, which is polypeptide. In this connection, it is of much interest to investigate the asymmetric reactions in which a synthetic polypeptide takes part, with respect to the effect of the primary and higher order structures. Such studies will not only serve as models for enzymes in order to throw light on the mechanism of their stereospecificity, but also open a way to develop specific catalysts for synthetic reactions.

1. Asymmetric Catalysis by Synthetic Polypeptides

Asymmetric reactions of synthetic polypeptides most closely related to enzymatic reaction are those actually catalyzed by synthetic polypeptide. Generally, asymmetric reactions may be divided into two categories:
1. asymmetric synthesis where an asymmetric center is newly formed in the product of one of the optical antipodes in excess, and
2. selection or differentiation of one of the optical antipodes of the reactant (optical resolution).

The studies of asymmetric synthesis have a rather long history in which the relation to the asymmetric character of living organisms has been the subject of interest (1). Among these studies, asymmetric syntheses catalyzed by synthetic polypeptide are considered most closely related, in a sense, to the stereospecific enzymatic reaction.

1.1. Asymmetric Hydrogenation Catalyzed by Metallic Complex of Synthetic Polypeptide

Akabori, Sakurai, Izumi and Fujii (2) were the forerunners in the asymmetric hydrogenation catalyzed by a metallic complex of a polypeptide, though not synthetic but naturally occurring silk fibroin. Palladium(II) chloride was adsorbed on silk fibroin of cultured silkworm and was reduced to metallic palladium, which was used as catalyst for the asymmetric hydrogenation to result in the formation of optically active α-amino acid. In the hydrogenation of diethyl α-acetoxyiminoglutarate [1], S-(+)-glutamic acid was obtained in 7.2% optical yield, while the hydrogenation of 4-benzylidene-2-methyloxazol-5-one [2] gave R-(+)-phenylalanine in 35.5% optical yield.

$$C_2H_5OCOCH_2CH_2\underset{\underset{N-OCOCH_3}{\overset{\|}{\underset{}{}}}}{\overset{}{-C}}-COOC_2H_5 \xrightarrow[\text{ii) HCl/H}_2O]{\overset{\text{i) H}_2,\ 96\ \text{kg/cm}^2}{\overset{50\ ^\circ C,\ CH_3OH}{}}} HOCO-CH_2CH_2-\overset{*}{\underset{NH_2}{CH}}-COOH \quad (1)$$

1

$$C_6H_5-CH=C-CO \xrightarrow[\text{ii) HCl/H}_2O]{\overset{\text{i) H}_2,\ 90\ \text{kg/cm}^2}{\overset{75\ ^\circ C,\ CH_3OH}{}}} C_6H_5-CH_2-\overset{*}{\underset{NH_2}{CH}}-COOH \quad (2)$$

2

Since silk fibroin of cultured silkworms contains relatively large amount of achiral glycine residue, silk fibroin of wild silkworms with a higher content of chiral alanine residue was examined as the carrier of palladium and optical yields of S-(+) 14.4% and R-(+) 32.1% were obtained in reactions (1) and (2), respectively (3). When amino groups of silk fibroin of cultured silk worm were acetylated and used as the carrier, the optical yield in reaction (2) was R-(+) 65,9%.

The active site of the catalyst has been considered to be metallic palladium deposited in the β-structure of silk fibroin as an asymmetric chelate. However, as a result of the examination of the reproducibility of these asymmetric hydrogenation reactions, the asymmetric modification ('poisoning') of palladium by S-α-amino acid was suspected, the latter being concomitantly formed by the hydrolysis of silk protein with hydrochloric acid from palladium(II) chloride in the course of the catalyst preparation (4).

Asymmetric hydrogenation catalyzed by palladium deposited on simpler synthetic polypeptides, homopolymers of α-amino acids or their derivatives, was studied by Beamer, Belding and Fickling (5, 6). In the hydrogenation in ethanol of α-methylcinnamic acid [3] and of α-acetamidocinnamic acid [4] to give 2-methyl-3-phenylpropanoic acid [5] and phenylalanine, respectively, the predominant optical antipode was found much dependent on the secondary conformation of the polypeptide (Table 1).

Table 1. Asymmetric hydrogenation of α-acetamidocinnamic acid catalyzed by palladium(II) on poly(amino acid) (5–6)

Poly(amino acid)	Chemical yield (%)	Optical yield (%)
Poly(γ-benzyl S-glutamate)	56.5	S 5.95
Poly(β-benzyl S-aspartate)	56.5	R 0.93
Poly(S-leucine)	92.3	S 5.16
Poly(S-valine)	68.0	R 4.25

$$\begin{array}{c} COOH \\ | \\ C-CH_3 \\ || \\ CH-C_6H_5 \end{array} \quad \xrightarrow{H_2} \quad \begin{array}{c} COOH \\ | \\ *CH-CH_3 \\ | \\ CH_2-C_6H_5 \end{array} \qquad (3)$$

$$\qquad\qquad 3 \qquad\qquad\qquad\qquad\qquad 5$$

$$\begin{array}{c} COOH \\ | \\ C-NH-COCH_3 \\ || \\ CH-C_6H_5 \end{array} \quad \xrightarrow[\text{ii) HCl/H}_2\text{O}]{\text{i) H}_2} \quad \begin{array}{c} COOH \\ | \\ *CH-NH_2 \\ | \\ CH_2-C_6H_5 \end{array} \qquad (4)$$

$$\qquad\qquad 4$$

When poly(S-valine) having the β-structure was used as the carrier of palladium, the products were dextro-rotatory, whereas the levorotatory products were obtained when poly(S-leucine) and poly(γ-benzyl S-glutamate) with right handed α-helical conformation were used. Furthermore, poly(β-benzyl S-aspartate) assuming left handed α-helical structure gave dextrorotatory products, in contrast to the polypeptides with right handed helices. The lower optical yield of the product obtained by the poly(β-benzyl S-aspartate) catalyst system was considered to be due to the less stable helical conformation of this polymer than the right handed α-helix of poly(γ-benzyl S-glutamate). Thus, these asymmetric hydrogenation reactions are affected more by the chirality of the secondary conformation of the polypeptide as the carrier, than by the chirality of the primary structure or the asymmetric center of amino acid residues.

Although palladium on polypeptide systems mentioned above are heterogeneous, homogeneous system may be realized by the complexation of ruthenium(III) ion to poly(S-glutamic acid), which was found effective as catalyst for the asymmetric hydrogenation of methyl acetoacetate (Hirai, Aikawa, Furuta (7)).

$$CH_3-\underset{\underset{O}{||}}{C}-CH_2-COOCH_3 \quad \xrightarrow{H_2} \quad CH_3-\overset{*}{\underset{\underset{OH}{|}}{CH}}-CH_2-COOCH_3 \qquad (5)$$

In this reaction, the optical yield of the product varies with the pH of the reaction mixture as shown in Fig. 1. Since poly(S-glutamic acid) in aqueous solution is known to assume right handed helical conformation below pH 4.5 while randomly coiled above pH 6.0, the above observation was considered to be related to such conformational change of poly(S-glutamic acid) — ruthenium complex with pH. In fact, upon the addition of ruthenium(III) ion the decrease in helical content with pH tended to become less noticeable (Fig. 2), suggesting the stabilization of the helical structure by the complexation, as also suggested in the copper(II) complex of poly(S-glutamic acid) (8). As seen in Fig. 3, the R-(−) antipode of the product was formed in excess by randomly coiled catalyst, while the yield of the S-(+) antipode increased with the increase in the helical content of the catalyst system. In the case of the monomeric S-glutamic acid — ruthenium(III) catalyst, metallic ruthenium precipitated in the early stage of the hydrogenation, the system being heterogeneous. Simi-

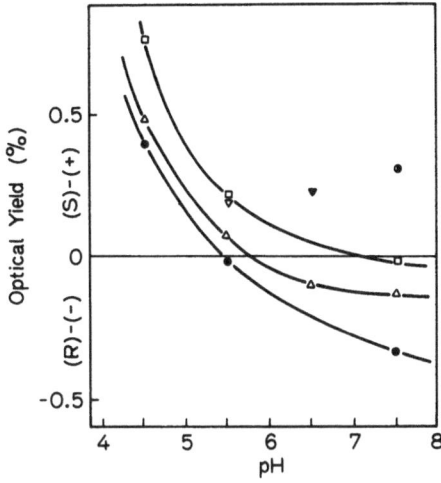

Fig. 1. Asymmetric hydrogenation of methyl acetoacetate catalyzed by ruthenium(III) complex with (7):
● Poly(S-glutamic acid), [COOH]/[Ru] = 10, △ [COOH]/[Ru] = 5, □ [COOH]/[Ru] = 2.5;
◐ Poly(R-glutamic acid), [COOH]/[Ru] = 10;
▼ S-glutamic acid, [COOH]/[Ru] = 5,
▽ [COOH]/[Ru] = 2.5

lar dependency of the optical yield of the product on the helical content of the poly(S-glutamic acid) — ruthenium(III) catalyst system was also found in the asymmetric hydrogenation of α-acetamidocinnamic acid [Eq. (4)] (9).

    In connection with these studies, the asymmetric hydrogenation catalyzed by Raney nickel modified with dipeptide should be mentioned. In the hydrogenation of methyl acetoacetate [Eq. (5)], Izumi and his co-workers (10) found that the

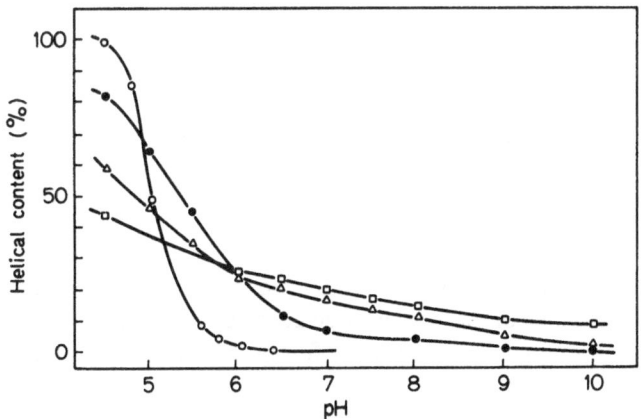

Fig. 2. Helical content of poly(R-glutamic acid) — ruthenium(III) complex in aqueous solution (7). $\overline{DP}$ = 100; [COOH]/[Ru], ○ no Ru, ● 10, △ 5, □ 2.5; room temperature; ionic strength 0.2

antipode excess in the product was determined by the chirality of the amino acid residue at the carboxyl terminal (Table 2). The catalytic activity of Raney nickel modified with α-amino acid or α-hydroxy acid in asymmetric hydrogenation has been studied widely (11).

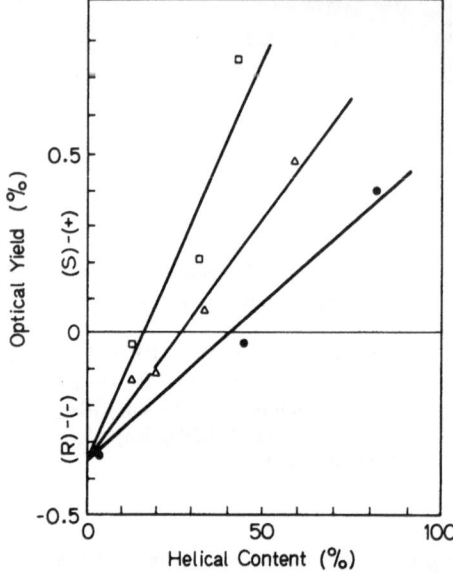

Fig. 3. Asymmetric hydrogenation of
methyl acetoacetate catalyzed by poly(S-
glutamic acid) – ruthenium(III) complex
(7). [COOH]/[Ru], ● 10, △ 5, □ 2.5

Table 2. Asymmetric hydrogenation of methyl
acetoacetate catalyzed by Raney nickel modified
with dipeptide and amino acid (*10*)

Modifier	Optical yield (%)
S-Leucine	R-(−) 5.26
S-Leucylglycine	S-(+) 5.35
Glycyl-S-leucine	R-(−) 1.96
S-Leucyl-S-leucine	R-(−) 2.01
S-Leucyl-R-leucine	S-(+) 5.63
S-Aspartic acid	R-(−) 5.26
Glycyl-S-aspartic acid	R-(−) 7.03

Typical biochemical hydrogenation (reduction) reactions are catalyzed by pros-
thetic enzymes with nicotinamide or flavin nucleotide as coenzyme. Model asymmetric hydrogenation of these types, however, await future studies.

1.2. Asymmetric Addition of Active Hydrogen Compound Towards Carbon-Carbon Double Bonds, Catalyzed by Synthetic Polypeptide

Addition reaction of active hydrogen compounds to carbon – carbon double bond
is one of the important classes of asymmetric synthesis in enzymatic reactions. Examples are the addition of water and ammonia to form S-malic acid and S-aspartic
acid, respectively.

$$H_2O + \quad \begin{array}{c} HOOC \\ \diagdown \\ \diagup \\ H \end{array} C=C \begin{array}{c} H \\ \diagup \\ \diagdown \\ COOH \end{array} \quad \xrightarrow{\text{fumarase}} \quad HOOC-CH_2-\overset{*}{C}H(OH)COOH \qquad (6)$$

$$NH_3 + \quad \begin{array}{c} HOOC \\ \diagdown \\ \diagup \\ H \end{array} C=C \begin{array}{c} H \\ \diagup \\ \diagdown \\ COOH \end{array} \quad \xrightarrow{\text{aspartase}} \quad HOOC-CH_2\overset{*}{C}H(NH_2)COOH \qquad (7)$$

As an asymmetric addition reaction of a similar type, Inoue and Ohashi (*12–15*) have studied the addition of thiol to α, β-unsaturated compounds, for example, fumaric ester (*15*), catalyzed by optically active polymeric and low molecular weight

$$CH_3(CH_2)_{11}SH + \quad \begin{array}{c} ROOC \\ \diagdown \\ \diagup \\ H \end{array} C=C \begin{array}{c} H \\ \diagup \\ \diagdown \\ COOR \end{array}$$

$$\xrightarrow{\text{Amine}^*} \quad CH_3(CH_2)_{11}S-\overset{*}{C}H(COOR)-CH_2COOR \qquad (8)$$

amines. As an extension of these studies, Fukushima and Inoue (*16*) examined the asymmetric addition of dodecanethiol to isopropenyl methyl ketone, affording (1-dodecylthiomethyl)ethyl methyl ketone, catalyzed by the amino end group of poly(γ-benzyl S-glutamate) (PBLG). PBLG with a primary amino group at the end

$$CH_3(CH_2)_{11}SH + CH_2 =\overset{\overset{\displaystyle CH_3}{|}}{C}-COCH_3 \xrightarrow{\text{PBLG}} CH_3(CH_2)_{11}S-CH_2-\overset{\overset{\displaystyle CH_3}{|}}{\underset{*}{C}}H-COCH_3$$

$$\text{PBLG: } H\{NH-\overset{*}{C}H-CO\}NH(CH_2)_3CH_3 \qquad (9)$$
$$\underset{\displaystyle CH_2CH_2COOCH_2C_6H_5}{|}$$

of the polymer chain may be conveniently obtained by the polymerization of the N-carboxy anhydride (NCA) of γ-benzyl S-glutamate, with butylamine as initiator. In this polymer, only the primary amino group has catalytic activity for the addition of the thiol to the unsaturated ketone. Furthermore, the number average degree of polymerization of PBLG thus formed is known to be almost equal to the mole ratio of monomer to initiator.

The catalytic activity of PBLG for this asymmetric addition reaction was examined for various mole ratios n of the NCA to initiator, butylamine, in the preparation of the PBLG in tetrahydrofuran. In the addition reaction carried out in chloroform, asymmetric synthesis was observed in every case, and the addition product having the largest absolute $[\alpha]_D$ value was obtained when the mole ratio n was 10 in the preparation of the catalyst (PBLG$_{10}$) (Table 3).

Although PBLG prepared by the polymerization of the NCA is not considered to be monodispersed (*17*), the results indicate that PBLG$_{10}$ contains a large amount

Table 3. Asymmetric addition of dodecanethiol to isopropenyl methyl
ketone catalyzed by poly(γ-benzyl S-glutamate) $(PBLG_n)^a$ (16)

Catalyst (g)		Chloroform (ml)	Time (day)	Conv. (%)	$[\alpha]_D$ of the product[b]
$PBLG_1$	(0.5)	20	20	64	−0.40
$PBLG_6$	(1.5)	30	26	66	−0.32
$PBLG_8$	(1.5)	30	30	50	−0.46
$PBLG_{10}$	(1.5)	30	37	68	−2.5
$PBLG_{12}$	(1.5)	30	44	50	−0.36
$PBLG_{14}$	(1.5)	30	48	50	−0.29
$PBLG_{20}$	(1.5)	30	48	74	−0.32
None		30	35	0	− −

[a] Reaction temp. 35 °C, dodecanethiol: 25 mmole, isopropenyl
   methyl ketone: 37 mmole, n: molar ratio of NCA to initiator in
   the preparation of PBLG.
[b] Measured in methanol (10 g/dl) at room temperature.

of molecules the amino end group of which is effective as catalyst for the asymmetric
synthesis. It is interesting to note that PBLG with a degree of polymerization > 8−10
may assume α-helical conformation (18). Since the primary structure around the
amino end group is identical in any $PBLG_n$, the results are considered to indicate
that such a conformational change of the catalyst with the molecular weight affects
the asymmetric synthesis. As to PBLG prepared from the NCA with $n > 10$, the
molecular weight distribution could be bimodal, the polymer being a mixture of a
larger amount of a fraction with a lower molecular weight (degree of polymerization
< 10) and a much smaller amount of a fraction with a higher molecular weight, as
reported for PBLG formed from NCA in dioxane (19) (though not in tetrahydrofuran). Thus, the low optical activity of the products obtained with $PBLG_n$ with
$n > 12$ might indicate the predominant role of the fraction with the low molecular
weight in the catalysis.

Differently from poly(γ-benzyl S-glutamate) (PBLG), poly(β-benzyl S-aspartate)
(PBLA) is known to assume a left-handed α-helical conformation which is less stable
than the α-helical conformation of PBLG (20). In this respect, it is of interest to examine the asymmetric addition of dodecanethiol to isopropenyl methyl ketone
[Eq. (9)] catalyzed by the terminal amino group of PBLA. By using PBLA, prepared
by the polymerization of the corresponding NCA in a mixture of 1,2-dichloroethane
and tetrahydrofuran with butylamine as initiator, Fukushima and Inoue (21) observed the occurrence of asymmetric synthesis as shown in Table 4.

All PBLA examined gave products having optical rotation of the same sign as
those obtained by PBLG as the catalyst. Therefore, the configuration of the predominant antipode of the product is determined by the chirality of the N-terminal unit of
the catalyst, but not by the handedness of the helices of these poly(amino acid)s.
However, each PBLA, especially when $n = 10$, gave a product with optical rotation
of smaller magnitude than that obtained by PBLG, prepared with the corresponding

Table 4. Asymmetric addition of dodecanethiol to isopropenyl methyl ketone catalyzed by poly(β-benzyl S-aspartate)[a] (21)

Mole ratio[b] n	Weight of catalyst (g)	Time (day)	Conv. (%)	$[\alpha]_D$ of the product[c]	$[\alpha]_D(PBLG_n)$[d] (16)
1[e]	0.7	17	53	−0.30	−0.40
5	1.0	26	91	−0.05	−0.32[f]
10	1.0	40	76	−0.48	−2.5
20	1.0	34	64	−0.24	−0.32

[a] Reaction temp., 35 °C; dodecanethiol, 25 mmole; isopropenyl methyl ketone,
   37 mmole; chloroform, 30 ml.
[b] Mole ratio of the NCA to initiator in the preparation of the catalyst.
[c] Measured in methanol (10 g/dl) at room temp.
[d] For the product obtained by poly(γ-benzyl S-glutamate) prepared with the
   corresponding n (PBLG$_n$).
[e] Reaction in 20 ml. chloroform.
[f] For $n = 6$.

NCA/initiator ratio n. This fact indicates that the stability of the α-helical conformation of these catalysts is important for the degree of asymmetric synthesis.

Furthermore, the same asymmetric addition reaction in chloroform was studied
by Ueyanagi and Inoue (22) using as catalyst the terminal amino group of poly(S-
alanine), insoluble in chloroform, prepared by the polymerization of the NCA initiated with butylamine in acetonitrile. As Table 5 shows, the optical rotation of the
addition product obtained by poly(S-alanine) as catalyst was larger than that obtained
by its terminal model, ethyl S-alaninate or S-alaninepropylamide, similarly to the
case of PBLG. When poly(S-alanine) was used as catalyst, however, the largest optical
rotation of the product was obtained when $n = 3$, n being the ratio of the NCA to the
initiator in the preparation of the polymer (Table 5).

Table 5. Asymmetric addition of dodecanethiol to isopropenyl
methyl ketone catalyzed by poly(S-alanine) (PLA$_n$)[a] and the
terminal models, in chloroform at 35 °C (22)

Catalyst	Time (days)	Conv. (%)	$[\alpha]_D$ of the product[b]
Ethyl S-alaninate	25	55	−0.34
S-Alaninepropylamide	40	86	−0.65
PLA$_3$	49	71	−1.94
PLA$_5$	41	64	−1.28
PLA$_8$	27	65	−0.91
PLA$_{10}$	41	67	−0.84
PLA$_{12}$	31	58	−0.66
PLA$_{20}$	58	72	−0.74

[a] n Represents the mole ratio of S-alanine NCA to initiator in
   the preparation of PLA$_n$.
[b] Measured in methanol at room temperature, 15 g/dl.

According to Komoto, Akaishi, Oya and Kawai (23), in the very early stages of the polymerization of S-alanine NCA in acetonitrile, the polymer formed ($\overline{DP}$ = 3–4) crystallizes in the form of antiparallel β-structure and precipitates. After this stage, the following several residues of the polymer assume random conformation, and after the polymer attains to $\overline{DP}$ = 8–10 the following residues grow assuming α-helical conformation. On the basis of these observations, for the asymmetric addition catalyzed by the amino end group of the insoluble poly(S-alanine) formed in acetonitrile, the β-structure of the polymer is considered the most effective.

Thus, in the asymmetric synthesis observed in the addition of dodecanethiol to isopropenyl methyl ketone catalyzed by the terminal amino group of poly(α-amino acid), the secondary conformation of the polymer catalyst plays an important role. The mechanistic pathway of this addition reaction is as follows (24, 25), in which

$$CH_3(CH_2)_{11}SH + \overset{|}{\underset{|}{N}}-* \rightleftarrows CH_3(CH_2)_{11}S^{-+}\overset{|}{\underset{|}{N}}H-* \tag{10}$$

$$CH_3(CH_2)_{11}S^{-+}\overset{|}{\underset{|}{N}}H-* + CH_2{=}\overset{\overset{\displaystyle CH_3}{|}}{\underset{\underset{\displaystyle COCH_3}{|}}{C}} \longrightarrow CH_3(CH_2)_{11}S-CH_2-\overset{\overset{\displaystyle CH_3}{|}}{\underset{\underset{\displaystyle COCH_3}{|}}{C}}{-}^{+}\overset{|}{\underset{|}{N}}H-* \tag{11}$$

$$CH_3(CH_2)_{11}S-CH_2-\overset{\overset{\displaystyle CH_3}{|}}{\underset{\underset{\displaystyle COCH_3}{|}}{C}}{-}^{+}\overset{|}{\underset{|}{N}}H-* \xrightarrow{\;(H^+)\;} CH_3(CH_2)_{11}S-CH_2-\overset{\overset{\displaystyle CH_3}{|}}{\underset{\underset{\displaystyle COCH_3}{|}}{C}}*H \quad + \tag{12}$$

$$+\overset{|}{\underset{|}{N}}-*$$

the configuration of the newly formed asymmetric carbon atom is determined at step (12), where the direction of the addition of the proton is asymmetrically controlled by the presence of the asymmetric counter cation, the protonated from of the terminal amino group of the polypeptide with asymmetric primary and secondary structures.

1.3. Selection of Optical Antipode in the Hydrolysis of Chiral Ester Catalyzed by Synthetic Polypeptide

Although the hydrolysis of esters catalyzed by synthetic polymers has been studied extensively, studies from the aspect of the stereospecificity have so far been rather limited.

Sheehan, Bennett and Schneider (26) prepared pentapeptides S-threonyl-S-alanyl-S-seryl-S-histidyl-S-aspartic acid 6 and S-seryl-γ-aminobutyryl-S-histidyl-γ-amino-butyryl-S-aspartic acid 7. These pentapeptides may be regarded as a model of α-chymotrypsin, having seryl and histidyl residues. When these peptides were used as catalyst for the hydrolysis of the optical antipodes of N-carbomethoxyphenylalanine p-nitrophenyl ester, the S antipode was found to be hydrolyzed faster than the R antipode by 1.5–2 times (Table 6). It is interesting to note that the catalytic rate was higher for pentapeptide 7 with flexible γ-aminobutyryl residues, while the relative specificity is higher for the more rigid pentapeptide 6.

Table 6. Hydrolysis of N-carbomethoxyphenylalanine p-nitrophenyl ester catalyzed by pentapeptide (26)

Optical antipode of substrate	Catalyst	k_2 (1/mol/min)
S	6	62
R	6	32
S	7	155
R	7	112
S or R	Imidazole	60

On the other hand, Imanishi, Tanihara, Sugihara and Higashimura (27) prepared the cyclic peptides cyclo(glycyl-S-histidyl) and cyclo(S-leucyl-S-histidyl) and examined the catalytic activity for the hydrolysis of the same ester as mentioned above. But no difference in rate was observed between the enantiomers of the substrate.

The catalytic activity of poly(S-lysine) – copper(II) complex in homogeneous aqueous solution for the hydrolysis of the optical antipodes of phenylalanine methyl ester was examined by Nozawa, Akimoto and Hatano (28). As shown in Table 7, the R antipode was hydrolyzed faster than the S antipode by 2–4 times, at pH = 7. At this pH poly(S-lysine) – copper(II) complex is in random conformation. At higher pH, where the complex assumes α-helical conformation, spontaneous hydrolysis would proceed with higher rate than the catalytic reaction.

Table 7. Hydrolysis of phenylalanine methyl ester catalyzed by poly(S-lysine)-copper(II) complex[a] (28)

$\overline{DP}$ of poly(S-lysine)	Ester antipode	Conc 10^3 (mol/l)	k (10^{-3} min^{-1})
470	R	8.04	6.13
	S	8.04	1.84
920	R	8.51	22.1
	S	9.22	9.12
	R	3.76	3.80
	S	4.26	0.92

[a] pH 7.0; 24.8 °C; [Cu], 5.2 x 10^{-4} mol/l; [Cu]/[N], 0.13.

The hydrolysis of an amino acid ester catalyzed by a copper complex is assumed to proceed through a coordination of the substrate to the complex followed by the hydrolysis of the coordinated substrate. The difference between the reactivities of

$$\text{Cu Complex} + \text{Ester} \; \overset{K}{\rightleftharpoons} \; (\text{Cu Complex})\,(\text{Ester}) \tag{13}$$

$$(\text{Cu Complex})\,(\text{Ester}) + \text{H}_2\text{Q} \; \overset{k}{\longrightarrow} \; (\text{Cu Complex})\,(\text{Acid}) + (\text{Alcohol}) \tag{14}$$

the entantiomeric substrates may be due to the difference in K and/or k, but no conclusion may be drawn from the data presented so far.

1.4. Selection of Optical Antipode in the Oxidation of Chiral Substrate Catalyzed by Metallic Complex of Synthetic Polypeptide

Prior to the hydrolysis of the ester mentioned above, Hatano, Nozawa, Ikeda and Yamamoto (29) studied the catalytic action of poly(S-lysine) (PLL) – copper(II) complex for the oxidation of 3,4-dihydroxyphenylalanine (DOPA) to the corresponding quinone.

$$\tag{15}$$

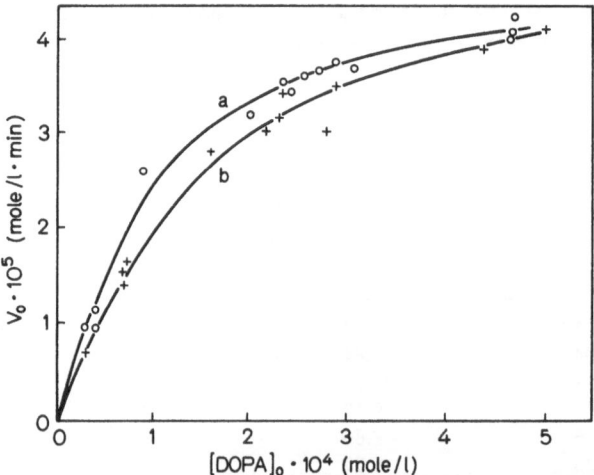

Fig. 4. Oxidation of DOPA catalyzed by poly(S-lysine) – copper(II) complex. pH = 10.5, 20 °C, $[Cu]/[N] = 0.13$, $[Cu] = 2.0 \times 10^{-4}$M; (a) RS-DOPA, (b) S-DOPA (29)

In the reaction at pH = 10.5, RS-DOPA was found to be oxidized more rapidly than S-DOPA, as seen in Fig. 4, whereas at pH = 6.9 no difference in the reactivities was observed between RS- and S-DOPA (Fig. 5). Poly(RS-lysine) − copper(II) complex showed no difference in catalytic activity towards RS- and S-DOPA (Fig. 6).

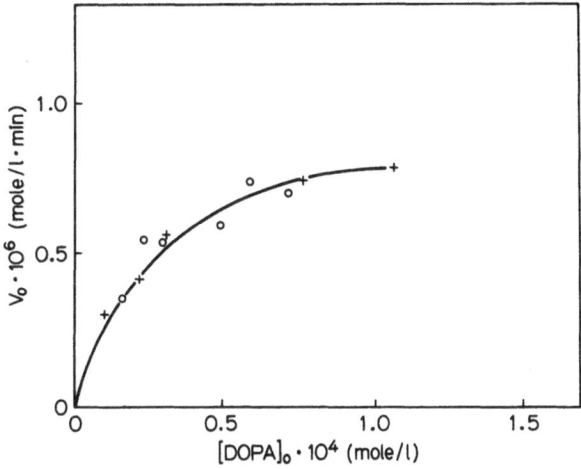

Fig. 5. Oxidation of DOPA catalyzed by poly(S-lysine) − copper(II) complex. pH = 6.9, 20 °C, [Cu]/[N] = 0.13, [Cu] = 2.0 x 10^{-4} M; o RS-DOPA, + S-DOPA (*29*)

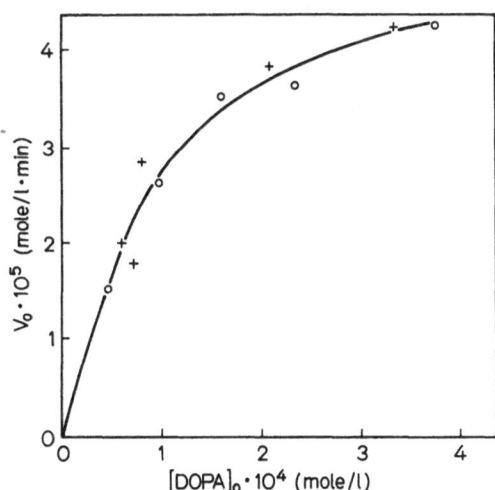

Fig. 6. Oxidation of DOPA catalyzed by poly(RS-lysine) − copper(II) complex. pH = 10.5, 20 °C, [Cu]/[N] = 0.13, [Cu] = 2.0 x 10^{-4}M; o RS-DOPA, + S-DOPA (*29*)

Furthermore, the initial rate of the oxidation showed a saturation at high substrate concentration (Fig. 4), characteristic of Michaelis − Menten type kinetics. Using the plot of the reciprocal initial rate versus the reciprocal substrate concentration (Fig. 7), the Michaelis constant (corresponding to the dissociation constant) of the catalyst — substrate complex and the rate constant of this complex may be calculated from the slope and the intercept of the line, respectively. As Fig. 7 and the results summarized in Table 8 show, the difference in the rate between S- and RS-DOPA is due to the difference in the stabilities of the catalyst — substrate complexes, but not to the difference in the reactivities of the complexes.

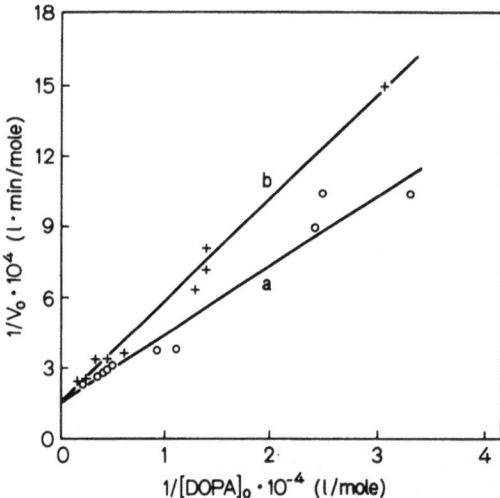

Fig. 7. Oxidation of DOPA catalyzed by poly(S-lysine) — copper(II) complex, reciprocal initial rate (Vo) versus reciprocal DOPA concentration. pH = 10.5, 20 °C, [Cu]/[N] = 0.13, [Cu] = 2.0×10^{-4}M; (a) RS-DOPA, (b) S-DOPA (29)

Table 8. Michaelis constant (K_m) and initial oxidation rate (V_0) in the oxidation of DOPA by poly(S-lysine) (PLL)-copper(II) and poly(RS-lysine) (PDLL)-copper(II) complexes[a] (29)

Catalyst	Substrate	pH	$V_0 \times 10^5$ (mol/l · min)	$K_m \times 10^4$ (mol/l)
PLL-Cu(II)	L-DOPA	10.5	1.04	2.25
PLL-Cu(II)	DL-DOPA	10.5	1.40	1.60
PDLL-Cu(II)	L-DOPA	10.5	1.60	1.68
PDLL-Cu(II)	DL-DOPA	10.5	1.60	1.68
PLL-Cu(II)	L-DOPA	6.9	0.067	0.34
PLL-Cu(II)	DL-DOPA	6.9	0.067	0.34

[a] For other conditions see Fig. 7.

Since poly(S-lysine) — copper(II) complex at pH = 10.5 assumes α-helical conformation while it is randomly coiled at pH = 6.9, the selective catalysis towards the entantiomeric substrates is considered to be related to the α-helical conformation of the catalyst. This was confirmed also by the comparison of the oxidation rates of R-DOPA and S-DOPA at various temperatures in relation to the α-helical content of the catalyst as obtained by the circular dichroic analysis. From these and other observations (30), a schematic model of the intermediate of the oxidation reaction has been proposed (Fig. 8) (31). In this bifunctional coordination of DOPA, the

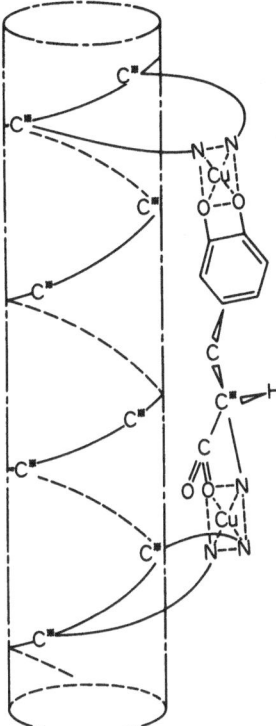

Fig. 8. A schematic model for the bifunctional coordination of S-DOPA to poly(S-lysine) – copper(II) complex (*31*)

catechol group is coordinated to a cupric ion to be activated, and the amino acid group also is coordinated to another cupric ion on the poly(S-lysine). In the latter coordination, R-DOPA is considered more favorable than S-DOPA, probably because of the lesser steric interaction between the β-methylene group and the right-handed α-helical chain of the polymer.

2. Selection of Optical Antipode in the Formation of Synthetic Polypeptide

In relation to the catalytic action of a macromolecular compound, the formation reaction of the macromolecular compound is important. The latter reaction is a 'catalytic' reaction at the end of macromolecule, in the sense that the reaction of the same substrate (monomer) is repeated at the growing end, retaining the same primary structure throughout the reaction.

When the conformation of the macromolecule affects the reactivity, an 'auto-catalytic' reaction would be observed. In this respect, the formation of poly(α-amino acid) by the polymerization of α-amino acid N-carboxy anhydride is the subject of interest, since the polymer being formed may assume specific conformations as α-helical, β, or randomly coiled.

2.1. Kinetic Studies of the Polymerization of α-Amino Acid N-Carboxy Anhydride

The polymerization of α-amino acid N-carboxy anhydride (NCA) is a well-known reaction for the preparation of poly(α-amino acid) with high molecular weight, which has played an important role as a model of protein (*32*).

$$x \quad \begin{array}{c} R \\ | \\ CH-CO \\ | \qquad\qquad O \\ NH-CO \end{array} \quad \xrightarrow{-x\ CO_2} \quad (-NH-\overset{\overset{\displaystyle R}{|}}{C}H-CO-)_x \qquad\qquad (16)$$

8

This reaction is initiated by various compounds such as amine, water, alcohol, metal alkoxide, organometallic compound etc. The reaction initiated by primary amine is one of the best studied, and proceeds by the repetition of the attack of primary amino group towards the 5-carbonyl group of NCA.

$$\begin{array}{c} R-CH-CO \\ |\ \ {}_{4}\ \ {}_{5} \\ |\ \ {}_{3}\ \ {}_{2} \\ NH-CO \end{array} \quad 1O + R'NH_2 \longrightarrow \quad \begin{array}{c} R-CH-CO-NHR' \\ |\qquad {}_{5} \\ |\qquad {}_{2} \\ NH-COOH \end{array} \qquad (17)$$

$$\xrightarrow{-CO_2} \quad \begin{array}{c} R-CH-CO-NHR' \\ |\qquad {}_{5} \\ NH_2 \end{array} \quad \xrightarrow{NCA} \quad R'NH\text{-}(CO-CHR-NH\text{-})_x H$$

In 1956, Doty and Lundberg (*18*) observed a rate increase in the middle of the polymerization of γ-benzyl S-glutamate NCA (8, $R = CH_2CH_2COOCH_2C_6H_5$) initiated by hexylamine in dioxane. As seen in Fig. 9, the polymerization exhibits a two stage kinetics, when the anhydride/initiator ratio (A/I) is high, the rate constant of the second stage being larger that that of the first stage by about five times. The conversion where such change in rate takes place is calculated to correspond to the degree of polymerization of the growing polymer of from 6 to 12. According to Blout and Asadourian (*33*), the polymer obtained in this reaction assumes α-helical conformation when A/I is higher than 50, while it forms the β-structure when A/I is lower than 4. Furthermore Goodman, Schmitt and Yphantis (*34*) prepared the oligmers of γ-methyl R-glutamate by a stepwise method, and found that the α-helical conformation of the polymer becomes stabilized at the degree of polymerization above 8–10. From these observations, Doty and his co-workers considered that the reaction of NCA with the growing end of the polymer having α-helical conformation is faster than that having randomly coiled conformation. The higher rate is considered to be due to the favorable orientation of NCA at the growing polymer end with

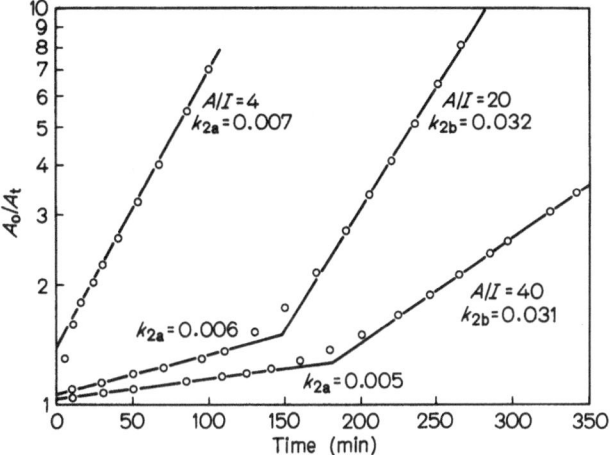

Fig. 9. Polymerization of γ-benzyl S-glutamate NCA by hexylamine in dioxane with various NCA/initiator ratios (A/I) (*18*)

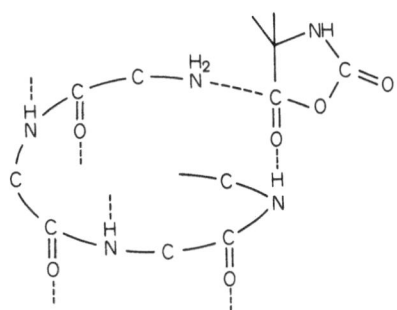

Fig. 10. A schematic model for the growing terminal in the polymerization of NCA by primary amine (*35*)

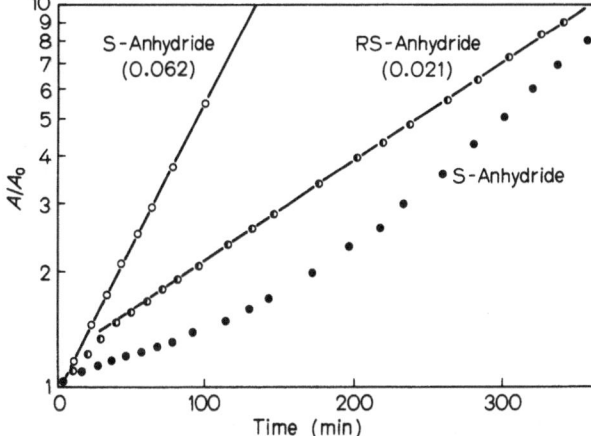

Fig. 11. Polymerization of γ-benzyl glutamate NCA initiated by S-polymer (*36*). ○ S-NCA, ◉ RS-NCA, ● R-NCA

α-helical conformation. A schematic model (Fig. 10) of the reaction was proposed by Weingarten (*35*).

Lundberg and Doty (*36*) further studied the polymerization of S-, RS- and R-NCA initiated by the terminal amino group of preformed S-polymer. As shown in Fig. 11, the polymerization of S—NCA proceeds with a high rate from the initial stage, while R-NCA polymerizes slowly with increasing rate at the later stage. The optical rotation of the reaction mixture changes with time, as shown in Fig. 12,

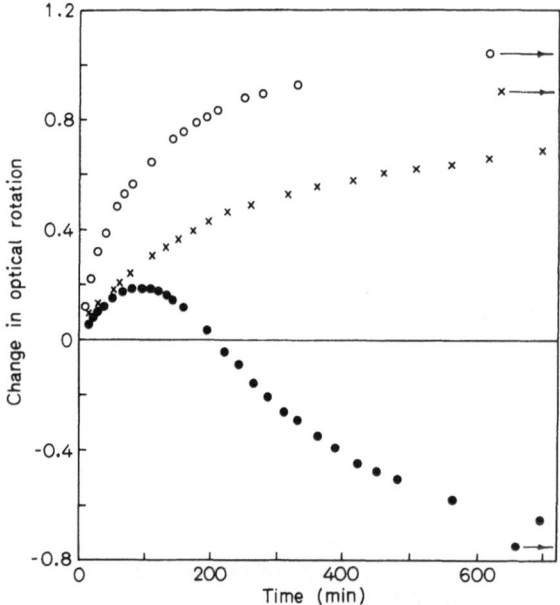

Fig. 12. Polymerization of γ-benzyl glutamate NCA initiated by S-polymer. Optical rotation of the reaction mixture (*36*). ○ S-NCA, x RS-NCA, ● R-NCA

taking as the standard the optical rotation when the reaction is started. S-NCA with negative rotation polymerizes to the polymer having positive rotation. In the polymerization of RS-NCA, it is interesting to note that the optical rotation of the reaction mixture does not return to its initial value even when the reaction is completed, indicating the incorporation of R-residues into the right handed helix. This is clearly observed in the initial stage of the polymerization of R-NCA, where the optical rotation becomes more positive although positive R-NCA should give negative polymer. Thus, the right handed α-helical conformation of S-polymer tends to be maintained even though R-antipode is incorporated. The rate decrease observed in the polymerization of the mixture of R- and S-NCAs (Fig. 13) (*36*) also corresponds to the facts mentioned above.

The rate increase in the course of the polymerization of γ-benzyl S-glutamate in dioxane (*37–39*) and the interpretation of the phenomenon have been a matter of controversy. Ballard and Bamford (*40*) observed the precipitation of the polymer

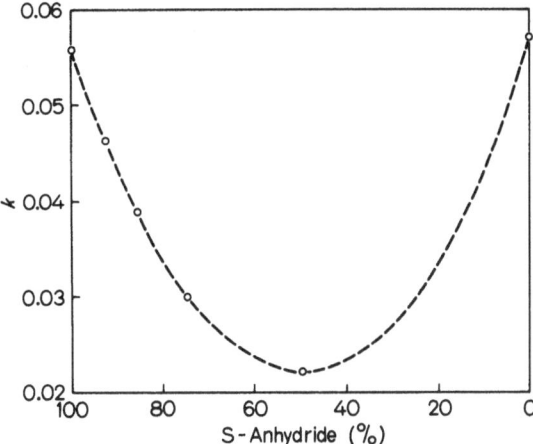

Fig. 13. Rate of polymerization of γ-benzyl glutamate NCA with various S-contents by hexylamine in dimethylformamide (*36*)

at the stage where the rate increase took place, and ascribed the rate increase to the adsorption of NCA to the polymer, enhancing the local concentration of NCA near the growing end. On the other hand, Nylund and Miller (*41*) observed the rate increase also in the polymerization in dimethylformamide, in which the reaction proceeds homogeneously at least in appearance.

More recently, Williams and Brown (*42–44*) made kinetic studies on the polymerization of the mixture of R and S antipodes of γ-benzyl glutamate NCA by hexylamine in dioxane. They also observed a two-stage reaction, and interpreted the phenomenon in terms of the adsorption of NCA to the polymer as a whole in the first stage, followed by the more specific adsorption of NCA to the three residues near the terminal of the growing α-helical polymer in the second stage. This explanation may be regarded as a combination of the proposed mechanisms by Ballard and Bamford, and by Doty and Lundberg.

2.2. Selection of Optical Antipode of N-Carboxy Anhydride in the Polymerization to Helix Forming Polypeptide

Although some of the kinetic data mentioned above indicate the selective reaction of the optical antipode of NCA in the polymerization, the direct evidence for this was obtained for the first time by Matsuura, Inoue and Tsuruta (*45–47*). In the polymerization of alanine NCA (*8*, R = CH_3) with the content of S antipode of 77.3%, initiated by methanol in tetrahydrofuran, the absolute optical rotation value of the polymer in trifluoracetic acid (TFA) increased with conversion through a maximum, then decreased (Fig. 14). This fact clearly indicates that the S antipode of the NCA polymerizes preferentially in the early stage, while in the later stage the residual R antipode polymerizes which becomes relatively rich in amount.

Although the polymerization of alanine NCA in tetrahydrofuran is a heterogeous reaction because of the insolubility of the polymer, a similar phenomenon is observed also in the polymerization of the mixture of R and S antipodes of γ-benzyl glutamate

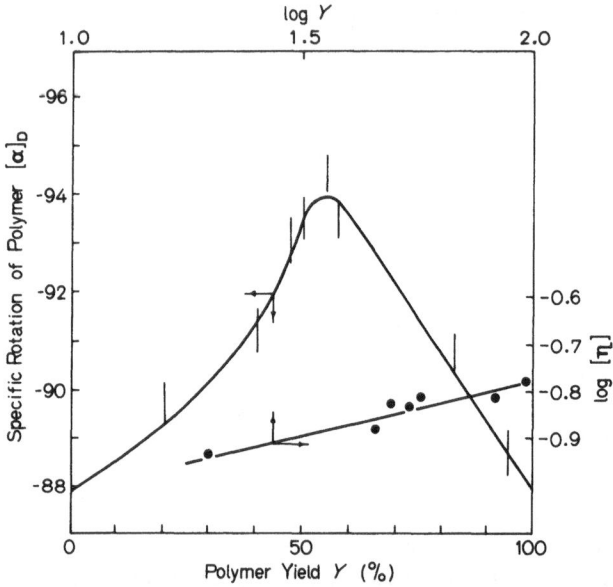

Fig. 14. Polymerization of S-77.3%-alanine NCA by methanol in tetrahydrofuran (*46*)

(8, R = $CH_2CH_2COOCH_2C_6H_5$) initiated by hexylamine in dimethylformamide, a homogeneous reaction (Tables 9 and 10) (*48*). In these tables, it is interesting to note that the α-helical content X_H of the polymer formed, as obtained by the analysis of the optical rotatory dispersion curve, increases throughout the reaction, indicating the incorporation of R antipode into the right handed helix of the polymer rich in S residues.

The conformation of the growing polymer in the above reaction may be studied directly by the circular dichroism measurement of the reaction mixture (*49*), as exemplified in Fig. 15 for the R and S copolymerization of γ-benzyl glutamate NCA initiated by butylamine in 1,2-dichloroethane. As seen in Fig. 15, the molar ellip-

Table 9. Polymerization of γ-benzyl S-55%-glutamate NCA initiated by butylamine[a] (*48*)

Time (h)	Polymer Conversion (%)	$[\alpha]_D^{25}$ (DMF)	$[\alpha]_D^{25}$ (TFA)	X_H (%)	$\overline{DP}$[b]
1	5.2	—	—	—	—
5	30.0	−13.1	—	26	31
9	52.8	−15.2	−5.6	34	40
15	75.6	−18.3	−6.9	37	49
42	94.4	−20.0	−5.1	42	55

[a] In dimethylformamide (DMF), A/I = 35, 26 °C.
[b] Viscosity average in dichloroacetic acid.

Table 10. Polymerization of γ-benzyl S-87%-glutamate NCA initiated by butylamine[a] (48)

No.	Time (h)	Polymer Conversion (%)	$[\alpha]_D^{25}$ (TFA)	X_H (%)	D 795/ D 2950	$\overline{DP}$[b]
1	3	40.4	−31.0	73	0.11	32
2	7	70.0	−32.2	85	0.32	55
3	10	80.9	−36.5	88	0.43	57
4	15	90.1	−33.9	93	0.41	65
5	23	95.6	−34.0	95	0.30	74

[a] In dimethylformamide (DMF), A/I = 40, 30 °C.
[b] Viscosity average in dichoroacetic acid.

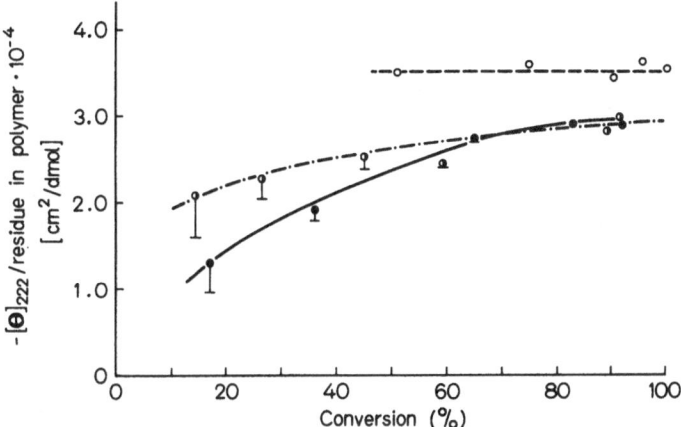

Fig. 15. Conversion vs. molar ellipticity per amino acid residue in polymer molecules in polymerization of γ-benzyl glutamate NCA by butylamine in 1,2-dichloroethane (49). ○ S-100%, $[A]_0/[I]_0 = 40$; ● S-65%, $[A]_0/[I]_0 = 40$; ◑ S-65%, $[A]_0/[I]_0 = 80$

ticity at 222 nm per amino acid residue in polymer, corresponding to the α-helical content, increased with conversion, in consistence with the data by the optical rotatory dispersion studies on the separated polymer (Tables 9 and 10).

The selective reaction of optical antipodes of NCA should reflect the stereoregularity or the sequential distribution of the antipodal residues in the polymer formed. Infrared spectral studies of the polymer have been known to provide useful information on the stereoregularity and the local conformation. As seen in Fig. 16 and Table 10 for the polymerization of γ-benzyl 87%-S-glutamate NCA in dimethylformamide (48), the strength of the absorption at 795 cm^{-1}, which has been reported to depend on the antipodal content of the polymer (50), changes with conversion through a maximum, corresponding to the observed selection of the antipode (Table 10).

Thus, the terminal amino group of the α-helical polymer, once formed, reacts preferentially with one of the antipodes of NCA. When the R to S ratio of NCA

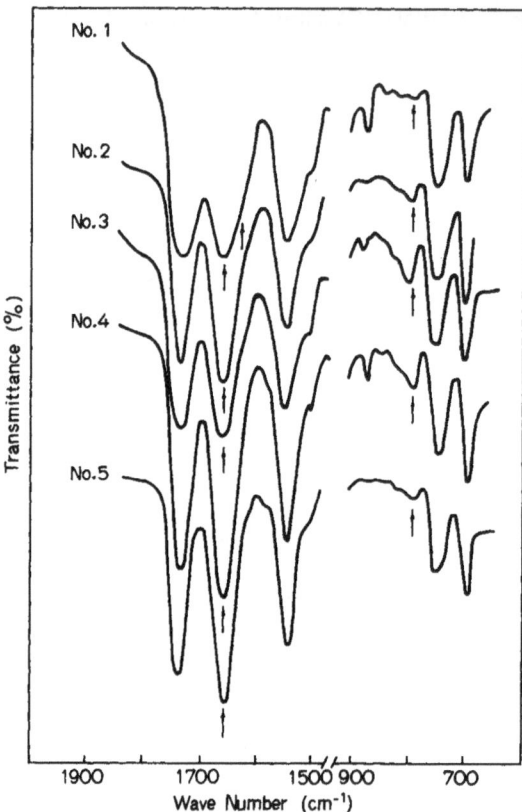

Fig. 16. Infrared spectra of poly(γ-benzyl glutamate) formed in polymerization of S-87%-NCA by hexylamine in dimethylformamide (48). Number of samples corresponds to that in Table 10

becomes close to one (racemic) the helix may not be formed, and the stereoregularity of the polymer formed is low, as evidenced by the fact that poly (γ-benzyl glutamate) obtained from the NCA with S content of 65% or 50% initiated by butylamine shows the infrared spectrum similar to that of the sequential copolymer such as $(RSSS)_x$ or $(RS)_x$ (Fig. 17) (51).

2.3. Polymerization of N-Carboxy Anhydride to Non-helix-forming Polypeptide

In the polymerization of the mixture of R and S antipodes of valine NCA (8, R = $CH(CH_3)_2$) initiated by butylamine in N,N-dimethylformamide and in 1,2-dichloroethane, the optical rotation of the polymer in trifluoroacetic acid scarcely varied at any conversion as shown in Fig. 18 (52), in clear contrast to the cases of alanine NCA and γ-benzyl glutamate NCA. Figure 18 implies that the content of antipodal residues of the polymer chains is constant throughout the reaction. As seen in Fig. 19, the plot of the optical rotation of the polymer against the S content of feed monomer gives a straight line, independently of the conversion, a fact which confirms that the S content of the polymers is equal to that of the feed monomers, even at half conversion.

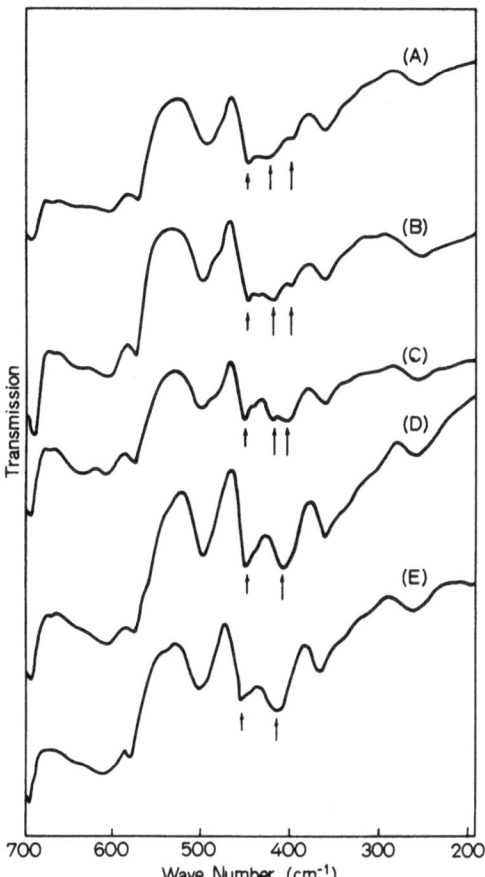

Fig. 17. Infrared spectra of poly(γ-benzyl R, S-glutamate) and sequential copolymers (51).
(A) S-50% polymer initiated by butylamine, A/I = 100. (B) (RSSS)$_x$ sequential copolymer. (C) (RS)$_x$ sequential copolymer. (D) S-50% polymer initiated by sodium methoxide, $\overline{DP}$= 151. (E) S-50% polymer initiated by sodium methoxide, $\overline{DP}$ = 98

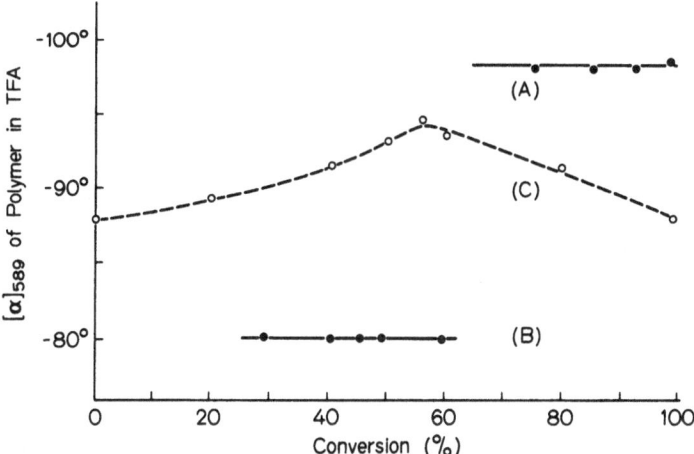

Fig. 18. Polymerization of valine NCA by butylamine (52)
(A) S-80% valine NCA in 1,2-dichloroethane
(B) S-75% valine NCA in dimethylformamide
(C) S-77.3% alanine NCA in tetrahydrofuran by methanol (46)

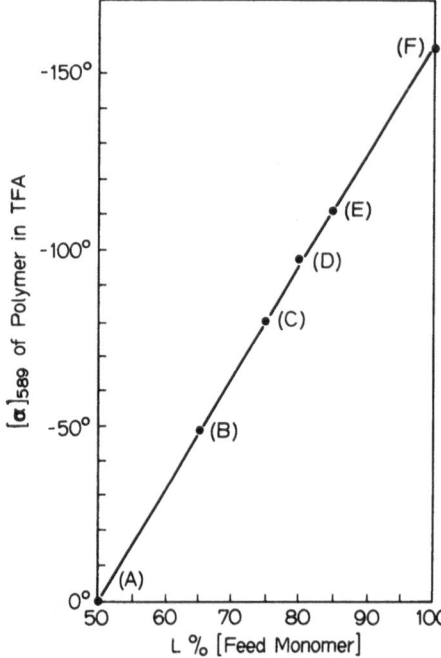

Fig. 19. Polymerization of valine NCA by butylamine. Optical rotation $[\alpha]_{589}$ of polymer in trifluoroacetic acid vs. S-content of feed monomer (52).
(A) S-50% (in dimethylformamide (DMF), conv. 52%). (B) S-65% (in DMF, conv. 48%). (C) S-75% (in DMF, conv. 29–60%). (D) S-80% (in CH_2Cl_2, conv. 76–100%). (E) S-85% (in DMF, conv. 52%). (F) S-100%, conv. 100%

As Makino, Inoue and Tsuruta (53) have reported, the ratio r of the reactivity of monomer (NCA) antipodes is generally expressed by the following equation:

$$\frac{d[R]}{d[S]} = \frac{[R]}{[S]} \cdot \frac{rF + 1}{r + F} \qquad (18)$$

where $d[R]/d[S]$ implies the ratio of the reaction rates of the two antipodes, $[R]/[S]$ the ratio of the concentration of the monomer antipodes, and F the ratio of the concentration of growing chain ends with R and S configurations, respectively. The results mentioned above mean

$$\frac{d[R]}{d[S]} = \frac{[R]}{[S]} \qquad (19)$$

throughout the reaction in the R and S copolymerization of valine NCA initiated by butylamine. Therefore, the comparison of the above two equations gives the following relationship.

$$\frac{rF + 1}{r + F} = 1 \qquad (20)$$

or $(r-1)(F-1) = 0$ (21)

In the polymerization of an α-amino acid NCA initiated by primary amine, it is known that the initiation reaction is much faster than the propagation reaction (54,

19), and forms a growing species with the same R/S ratio as the feed monomer. Accordingly, it is impossible that F, the ratio of the concentrations of R and S growing species, could be 1 throughout the reaction, except for the polymerization of racemic NCA. r, the ratio of the reactivities of monomer antipodes, is therefore equal to 1. Thus, in the R and S copolymerization of valine NCA, selection of monomer antipodes does not take place, and the possibility that the correspondence of the S-content of the polymer to that of the feed monomer might be due to the complete stereoselection, resulting in a mixture of all-R-polymers and all-S-polymers with the same ratio as the R/S ratio of the feed monomer, is excluded.

This phenomenon is in remarkable contrast to the cases of alanine NCA and γ-glutamate NCA, where a selection of monomer antipodes by α-helical polymer chain is observed. The lack of selectivity of the growing poly(valine) chains cannot be attributed to the heterogeneity of the reaction mixture due to the insolubility of the polymer, since the polymerization of alanine NCA in tetrahydrofuran does demonstrate the selectivity, as mentioned previously, despite the heterogeneous reaction observed. The determining factor seems to be the different conformation acquired by the growing poly(valine) as compared to the α-helices of poly(γ-benzyl glutamate) and poly(alanine), both in solution and solid state.

Poly(S-valine) has been thought not to be able to form an α-helix but have a β-structure (*55–57*). By the infrared spectral studies of copoly(R, S-valine)s obtained in the polymerization of the NCA, the antiparallel β-structure of the (S)-polymer is found to be substantially retained even when R residues are incorporated, as evidenced, for example, by the fact that an amide 1 peak at 1630 cm^{-1} and a shoulder peak at 1690 cm^{-1} are almost unchanged in the polymers with S contents from 100% to 50% (*58*). The same conclusion may be obtained from X-ray powder diffraction studies of these polymers (*58*). The diffraction pattern of poly(S-valine) has two prominent peaks which correspond to the interchain spacing 4.67 Å and the intersheet spacing 9.51 Å, of the β-pleated sheet structure, respectively. With the decrease in the S content of the polymer, the interchain spacings are almost unchanged, whereas the intersheet spacing increases gradually (Table 11). Thus, the β-structure is substantially retained at the expense of the increase in the intersheet spacing. These facts are considered to be related to the absence of selection of the optical antipode in the R and S copolymerization of valine NCA.

Table 11. X-ray powder diffraction pattern of poly(R, S-valine) obtained by the polymerization of the NCA[a] (*2, 58*)

S content (%)	Conversion (%)	Intersheet spacing, A (diffraction angle 2θ)		Interchain spacing, A (diffraction angle 2θ)	
100	57	9.51	(9.3°)	4.67	(19.0°)
85	52	9.61	(9.2°)	4.67	(19.0°)
75	60	9.94	(8.9°)	4.67	(19.0°)
65	48	10.05	(8.8°)	4.67	(19.0°)
50	52	9.94	(8.9°)	4.58	(19.4°)

[a] CuKα-ray, λ = 1.5418 A.

According to Imanishi, Kugimiya and Higashimura (59), no difference in the reactivities is observed between the optical antipodes of phenylalanine NCA (8, R = $CH_2C_6H_5$) in the polymerization initiated by the terminal secondary amino group of poly(N-methyl-S-alanine) (9). Although poly-(N-methyl-S-alanine) is considered

$$(C_2H_5)_2N{\small\{}CO-\underset{*}{CH}-N{\small\}}_xH$$

with CH_3 groups attached.

9

to assume some specific secondary structure, it is different from the α-helical conformation of N-unsubstituted poly(α-amino acid).

2.4. Other Problems Concerning the Stereochemistry of Polymerization of N-Carboxy Anhydride

For the selection of the NCA antipodes, it is, of course, the local conformation around the growing terminal that directly plays the important role. Therefore, the polymerizations caused by different types of initiator, having growing terminals of different chemical structure, would exhibit different selectivities with respect to the antipodal monomers. In fact, for example, the R and S copolymerization of γ-benzyl glutamate NCA initiated by sodium methoxide proceeds more selectively than that by butylamine (49, 51) (Fig. 17). In the R and S copolymerization of β-benzyl aspartate NCA, on the other hand, the stereoregularities of the polymers obtained by the above two initiators are very similar to each other (60). Since the left handed α-helical conformation of poly(β-benzyl S-aspartate) is less stable than the right handed helix of poly(γ-benzyl S-glutamate), the high selectivity of sodium methoxide initiated system is considered to become explicit when the polymer being formed assumes stable α-helical conformation.

Although it could be argued that the selection of the monomer antipode would be due to the selective adsorption to the polymer chain as a whole, the fact that the selectivities are different depending upon the initiator (or the chemical mechanism) in the formation of the same poly (amino acid), as described above and in another paper (53), excludes the importance of such adsorption. The selection of the monomer antipode was reported (61) also in the R and S copolymerization of an activated amino acid as below

Thus, in the polymerization of α-amino acid N-carboxy anhydride (NCA) initiated by primary amine, the α-helical conformation of the polymer being formed is con-

$$EtO-\underset{\underset{O}{\|}}{C}-S-\underset{\underset{S}{\|}}{C}-NH-\underset{*}{CH}-COOH \rightarrow {\small\{}NH-\underset{*}{CH}-CO{\small\}}_x \qquad (22)$$

with R groups attached to the CH positions.

sidered to be an important factor in the different reactivities of the optical antipodes of NCA at the terminal amino group of the growing polymer. Such a conformational effect is considered to be closely related to the conformational effect of poly (amino acid) catalyst in the asymmetric syntheses described previously. Furthermore, such an 'autocatalytic' effect in the formation of polypeptide would be related to the amplification of asymmetry in the course of chemical evolution from prebiotic organic compound with low molecular weight to polymeric compound (62, 63).

3. Asymmetric Reaction on Polypeptide Chain

As to the origin and the evolution of the asymmetry of organic compounds, there have been presented various hypotheses. Concerning the formation of optically active polymer in chemical evolution, three fundamental types would be expected: 1. the formation from optically active compounds of low molecular weight (monomer), 2. asymmetric reaction in the formation from optically inactive monomer and 3. asymmetric reaction on optically inactive preformed polymer. Among these, the theory presented by Akabori (64, 65) is of the third type mentioned above. The theory proposes the formation of poly(glycine) without starting from amino acid, followed by the reaction of active methylene group of poly(glycine) with aldehyde etc. under some asymmetric conditions to form optically active amino acid residues with a substituent.

In this connection, Sperling and Elad (66) observed asymmetric induction in the photo-alkylation by ultra-violet light of glycine residue in peptide containing glycine and optically active amino acid residues (Table 12). In this case as well, the secondary conformation of polypeptide appears to exhibit a remarkable influence, as shown

Table 12. Asymmetric induction in the photo-alkylation of peptide (66)

Peptide	Alkylating agent	Amino acid formed			
		Norleucine		Phenylalanine	
		R(%)	S(%)	R(%)	S(%)
Tfa-Gly-S-Ala-OMe	l-Butene	46	54		
Tfa-Gly-S-Ala-OMe	Toluene			47.5	52.5
Tfa-Gly-S-Leu-OMe	l-Butene	41	59		
Tfa-S-Ala-Gly-S-Ala-OMe	Toluene			60	40
Tfa-S-Pro-Gly-S-Pro-OMe	l-Butene	52	48		
Tfa-S-Pro-Gly-S-Pro-OMe	Toluene			55	45
[S-Ala-Gly-S-Ala]$_n$	Toluene			30	70
[S-Pro-Gly-S-Pro]$_n$	l-Butene	60	40		
[S-Pro-Gly-S-Pro]$_n$	Toluene			67.5	32.5

$$[-NH-CH_2-CO-NH-\overset{\overset{\textstyle R}{|}}{C}H-CO-] \xrightarrow[\text{light}]{\text{olefin}} [-NH-\overset{\overset{\textstyle R'}{|}}{C}H-CO-NH-\overset{\overset{\textstyle R}{|}}{C}H-CO-] \tag{23}$$

by the facts that the polypeptide [S–Pro–Gyl–S–Pro]$_n$ gives higher optical yields than the corresponding tripeptides, and that the polypeptide [S–Ala–Gly–S–Ala]$_n$ gives the excess enantiomer which is different from that obtained in the corresponding tripeptide. A similar asymmetric photo-alkylation was observed also in the reaction induced by visible light (67).

4. Conclusion

As shown in the catalytic reactions, the formation reactions and others, the asymmetric reactions of synthetic polypeptides are characterized by the participation of specific secondary structures. Such influence of the secondary conformation on the asymmetric reaction is considered to have a general importance in relation not only to the stereospecificity of the reaction of enzyme, but also to a possible way by which present-day enzymes obtained their high stereospecificity in the course of chemical evolution.

Acknowledgement. The author expresses sincere gratitude to Professor Teiji Tsuruta for his constant encouragement.

5. References

1. Morrison, J. D., Mosher, H. S.: Asymmetric organic reactions. Prentice-Hall 1971.
2. Akabori, S., Sakurai, S., Izumi, Y., Fujii, Y.: Nature **178**, 323 (1956).
3. Akabori, S., Izumi, Y., Fujii, Y.: Nippon Kagaku Zasshi, **78**, 886 (1957).
4. Izumi, Y.: Kagaku to Kogyo **20**, 1354 (1967).
5. Beamer, R. L., Belding, R. H., Fickling, C. S.: J. Pharm. Sci. **58**, 1142 (1969).
6. Beamer, R. L., Belding, R. H., Fickling, C. S.: J. Pharm. Sci. **58**, 1419 (1969).
7. Hirai, H., Akikawa, Y., Furuta, T.: Fall Meeting of Chem. Soc. Japan (Aug. 1970, Sapporo), 30B10.
8. Takesada, H., Yamazaki, H., Wada, A.: Biopolymers **4**, 713 (1966).
9. Hirai, H., Shindo, N., Hashimoto, R.: 27th Ann. Meeting of Chem. Soc. Japan (Oct. 1972, Nagoya) 4E07.
10. Izumi, Y., Tatsumi, S., Imaida, M., Okubo, K.: Bull. Chem. Soc. Japan **43**, 566 (1970).
11. Izumi, Y.: Angew. Chem. Int. Ed. **10**, 871 (1971).
12. Inoue, S., Ohashi, S., Tabata, A., Tsuruta, T.: Makromol. Chem. **112**, 66 (1968).
13. Ohashi, S., Inoue, S.: Makromol. Chem. **150**, 105 (1971).
14. Ohashi, S., Inoue, S.: Makromol. Chem. **160**, 69 (1972).
15. Inoue, S., Ohashi, S., Unno, Y.: Polymer J. **3**, 611 (1972).
16. Fukushima, H., Ohashi, S., Inoue, S.: Makromol. Chem. **176**, 2751 (1975).
17. Mitchell, J. C., Woodward, A. E., Doty, P.: J. Amer. Chem. Soc. **79**, 3955 (1957).
18. Doty, P., Lundberg, R. D.: J. Amer. Chem. Soc. **78**, 4810 (1956).
19. Cosani, A., Peggion, E., Scoffone, E., Verdini, A. S.: Makromol. Chem. **97**, 113 (1966).
20. Fasman, G. D., in: Polyamino acids (ed. G. D. Fasman), p. 499. New York: Marcel Dekker 1967.
21. Fukushima, H., Inoue, S.: Makromol. Chem. **176**, 3609 (1975).
22. Ueyanagi, K., Inoue, S.: Makromol. Chem., in press.
23. Komoto, T., Akaishi, T., Oya, M., Kawai, T.: Makromol. Chem. **154**, 151 (1972).
24. Mallik, K. L., Das, M. N.: Z. Phys. Chem. **41**, 35 (1964).
25. Dmuchovsky, B., Vineyard, B. D., Zienty, F. B.: J. Amer. Chem. Soc. **86**, 2874 (1964).
26. Sheehan, J. C., Bennett, G. B., Schneider, J. A.: J. Amer. Chem. Soc. **88**, 3455 (1966).
27. Imanishi, Y., Tanihara, M., Sugihara, T., Higashimura, T.: 23rd Ann. Meeting of Soc. Polymer Sci. Japan (1974, Tokyo), 5A06.
28. Nozawa, T., Akimoto, Y., Hatano, M.: Makromol. Chem. **158**, 21 (1972).
29. Hatano, M., Nozawa, T., Ikeda, S., Yamamoto, T.: Makromol. Chem. **141**, 11 (1971).
30. Hatano, M., Nozawa, T.: Progress in Polymer Science Japan **4**, 223 (1972).
31. Nozawa, T., Hatano, M.: Makromol. Chem. **141**, 31 (1971).
32. As a review on the polymerization of NCA, see: Shalitin, Y. in: Ring-opening polymerization (ed. K. C. Frisch and S. L. Reegen), p. 421. New York: Marcel Dekker 1969.
33. Blout, E. R., Asadourian, J.: J. Amer. Chem. Soc. **78**, 955 (1956).
34. Goodman, M., Schmitt, E. E., Yphantis, D. A.: J. Amer. Chem. Soc. **84**, 1228 (1962).
35. Weingarten, H.: J. Am. Chem. Soc. **80**, 352 (1958).
36. Lundberg, R. D., Doty, P.: J. Am. Chem. Soc. **79**, 3961 (1957).
37. Ballard, D. G. H., Bamford, C. H.: J. Am. Chem. Soc. **79**, 2336 (1957).
38. Doty, P., Lundberg, R. D.: J. Am. Chem. Soc. **79**, 2338 (1957).
39. Ballard, D. G. H., Bamford, C. H., Elliott, A.: Makromol. Chem. **35**, 222 (1960).
40. Ballard, D. G. H., Bamford, C. H.: J. Chem. Soc. **1959**, 1039.
41. Nylund, R. E., Miller, W. G.: Biopolymers **2**, 131 (1964).
42. Williams, F. D., Eshaque, M., Brown, R. D.: Biopolymers **10**, 753 (1971).
43. Williams, F. D., Brown, R. D.: Makromol. Chem. **169**, 191 (1973).
44. Williams, F. D., Brown, R. D.: Biopolymers **12**, 647 (1973).
45. Matsuura, K., Inoue, S., Tsuruta, T.: Makromol. Chem. **85**, 284 (1965).
46. Tsuruta, T., Inoue, S., Matsuura, K.: Biopolymers **5**, 313 (1967).
47. Matsuura, K., Inoue, S., Tsuruta, T.: Kogyo Kagaku Zasshi **69**, 127 (1966).
48. Inoue, S., Matsuura, K., Tsuruta, T.: J. Polymer Sci. C **23**, 721 (1968).

49. Akaike, T., Makino, T., Inoue, S., Tsuruta, T.: Biopolymers 13, 129 (1974).
50. Tsuboi, M., Mitsui, Y., Wada, A., Miyazawa, T., Nagashima, N.: Biopolymers 1, 297 (1963).
51. Akaike, T., Inoue, S., Itoh, K.: Biopolymers 13, 1713 (1974).
52. Akaike, T., Aogaki, Y., Inoue, S.: Biopolymers 14, 2577 (1975).
53. Makino, T., Inoue, S., Tsuruta, T.: Makromol. Chem. 150, 137 (1971).
54. Bamford, C. H., Block, G., in: Polyamino acids, polypeptides and proteins, (ed. M. A. Stahmann) pp. 65–78. Madison: Wisconsin Press 1962.
55. Blout, E. R., de Loze, C., Bloom, S. M., Fasman, G. D.: J. Amer. Chem. Soc. 82, 3787 (1960).
56. Bloom, S. M., Fasman, G. D., de Loze, C., Blout, E. R.: J. Amer. Chem. Soc. 84, 458 (1962).
57. Fraser, D. B., Harrap, B. S., MacRae, T. P., Stewart, G. H., Suzuki, E.: J. Mol. Biol. 12, 482 (1965).
58. Akaike, T., Inoue, S., Itoh, K.: Biopolymers, in press.
59. Imanishi, Y., Kugimiya, K., Higashimura, T.: Biopolymers 13, 1205 (1974).
60. Akaike, T., Inoue, S.: to be published.
61. Higashimura, T., Suzuoki, K., Kato, H., Okamura, S.: Makromol. Chem. 108, 129 (1967).
62. Calvin, M.: Chemical evolution, molecular evolution towards the origin of living systems on the earth and elsewhere, pp. 170–171. Oxford: University Press 1969.
63. Inoue, S.: Viva Origino 2, 21 (1973).
64. Akabori, S.: Kagaku 25, 54 (1955).
65. Akabori, S.: The Origin of Life on the Earth, I.U.B. Symposium Series, 1, Pergamon Press (1959) p. 189.
66. Sperling, J., Elad, D.: J. Amer. Chem. Soc. 93, 967 (1971).
67. Schwarzberg, M., Sperling, J., Elad, D.: J. Amer. Chem. Soc. 95, 6418 (1973).

Received January 23, 1976

Study of Polymers by Inverse Gas Chromatography

Jean-Marie Braun and James E. Guillet

Department of Chemistry, University of Toronto, Toronto, Canada M5S 1A1

Table of Contents

I. Introduction

The application of gas chromatography to the study of polymers has long been hampered by the negligible volatility of polymeric materials. As a result investigations were restricted to the analysis of volatiles in polymeric compounds such as residual monomers or additives. An alternative solution was developed, whereby the volatility of the polymer samples was "enhanced" by pyrolysis. The resulting low molecular weight fragments were then analyzed by gas chromatography and information on the structure and composition of the polymer could be derived.

The direct approach to the study of polymer properties and interactions by gas chromatography was initiated by the work of Smidsrød and Guillet (1) on poly(N-isopropyl acrylamide). The polymer was used as the stationary phase and its inter-action with known, volatile solutes was recorded. The name "inverse gas chromatography" is customarily given to this technique. From the magnitude and temperature dependence of the interaction, properties of the polymer-solute solution as well as those of the pure polymer can be measured. The facility of the gas chromatographic route to study highly concentrated polymer solutions has represented a major incentive to the rapid development of this method.

The necessary apparatus is simple, consisting of a short column packed with an inactive material containing the polymer dispersed as a thin film on the surface. In some cases the column may also be packed directly with the polymer in either film, fiber, or powder form. A uniform flow of inert gas is maintained through the column, and a pulse of probe molecules is injected at one end and detected at the other by a suitable detector. A small pulse of noninteracting gas can also be injected with the probe molecules to aid in detection of the carrier gas front. The usual data recorded are the temperature and pressure drop across the column and the time of the peak maxima.

II. Inverse Gas Chromatography

Gas chromatography (GC) is based on the distribution (partition) of a volatile solute between a mobile gas phase and a stationary liquid (GLC) or solid (GSC). The solute is introduced at the column inlet as a plug of vapor and is swept through the column by the carrier gas. Depending on the partition coefficients, differing volumes of carrier gas are required to elute the solutes from the column. Owing to the dynamics of the experiment, the results are usually expressed in GC as retention volumes rather than partition coefficients. For convenience, solute retention data are given either at column temperature or at 0 °C, with

$$V_g = (273.2/T) \, (V_N/w) = (273.2/T)K \tag{1}$$

where $V_g (cm^3 g^{-1})$ is the specific retention volume corrected to 0 °C, V_N (cm^3) is the net retention volume at temperature T, w is the mass of stationary phase in the column, and K is the partition coefficient. For a more detailed discussion of the theory and experimental requirements of GC the reader is referred to the many excellent texts and reviews on the subject (2–5).

The variation of the retention volume with temperature is such that a plot of the logarithm of the retention volume versus reciprocal of the absolute temperature, called a retention diagram, is linear. The slope of this straight line is related to the enthalpy of the process, solution in the stationary phase (GLC) or adsorption on the surface (GSC), with,

$$\frac{\partial \ln V_g}{\partial (1/T)} = -\Delta H/R \tag{2}$$

where $-\Delta H$ is the corresponding enthalpy and R the gas constant.

Early studies on polymer stationary phases revealed that retention diagrams exhibited certain singularities such as pictured in Fig. 1. They were found to correspond

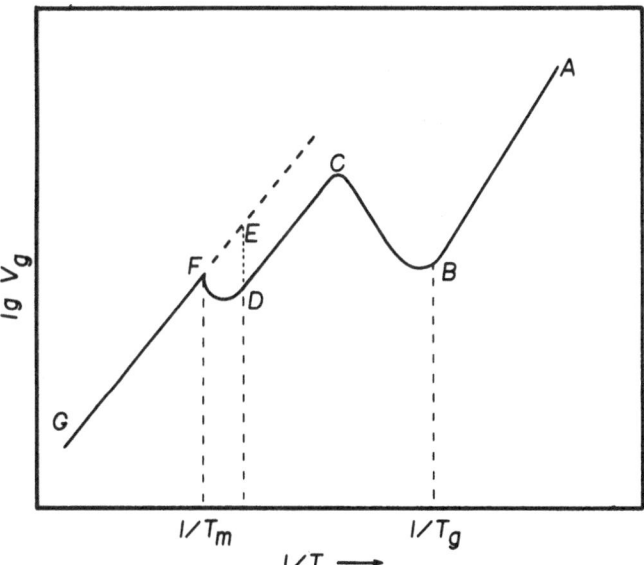

Fig. 1. Generalized retention diagram for semicrystalline polymer

to transitions of the polymer stationary phase. Both glass (1) and melting (6, 7) transitions were thus detected. These transitions of the stationary phase are of paramount importance in inverse gas chromatography for they determine the nature of the information that can be derived from any one experiment (8, 9). In conventional solution thermodynamics the polymer is usually designated as the solute and the small molecule is the solvent. However, in the GC experiment the polymer is the major phase present. To avoid confusion with solution terminology, GC studies are often referred to as "molecular probe" experiments, the probe being the volatile substance (solute).

In the temperature region corresponding to segment AB of Fig. 1, the polymer is below its glass transition temperature (T_g) and penetration of the solute molecules into the bulk of the polymer phase is precluded. Retention proceeds exclusively by surface adsorption and the corresponding retention diagram is linear. Information on the surface properties of the polymer can be obtained in this temperature range. At point B, corresponding to the glass transition, penetration of the solute into the bulk of the polymer begins, causing an increase of retention volume with temperature. Due to an initially slow rate of diffusion of the solute into and out of the stationary phase, nonequilibrium conditions prevail. As the temperature is increased in region BC the diffusion coefficient rises sharply, leading to equilibrium conditions at point C.

At temperatures below the melting point of the polymer, in region CD, retention proceeds by bulk sorption but the polymer-solute interaction is restricted to the amorphous domains of the stationary phase. Upon melting, in region DF, the fraction of amorphous material increases, leading to an increase in retention volume. At temperatures above the melting point, segment FG, a linear retention diagram, corresponding to bulk sorption into the completely amorphous polymer, is obtained. By extrapolation of this line to lower temperatures (dashed line FE), the crystalline content of the stationary phase can be determined by comparison of the experimental retention volume with the extrapolated value.

In region FG the polymer is completely amorphous and the properties of the polymer-solute solution can be investigated. Care must be taken to ensure that any contribution from surface adsorption is taken into account. Usually for thick polymer films contributions from surface adsorption will be small. For noncrystalline polymers, liquid-like behavior is observed from point C onwards. The location of point C on the temperature axis depends both on the polymer-solute system considered and on experimental conditions, film thickness and flow rate. For most polymers, equilibrium bulk sorption is achieved at temperatures in excess of about $T_g + 50°$.

III. The Glass Transition and Retention Mechanisms

Reversal from the normal chromatographic behavior at T_g was first detected by Smidsrød and Guillet (1) with poly(N-isopropyl acrylamide) stationary phases. By comparing retention data on columns of different loading it was shown that retention changed from surface adsorption below T_g to bulk sorption above T_g. The change in retention mechanism at T_g was attributed to an increased molecular mobility of the polymer segments at and above T_g, allowing for the penetration of the solute molecules into the bulk of the polymer. Guillet and collaborators (10–13) later detected the transitions of poly(methyl methacrylate) (10, 12), poly(vinyl chloride) (10), polystyrene (10, 13), and poly(acrylonitrile) (11) stationary phases. The reliability of the method was further tested with polystyrenes of varying molecular weights, poly(methyl methacrylate)s of differing tacticity and copolymers of buta-

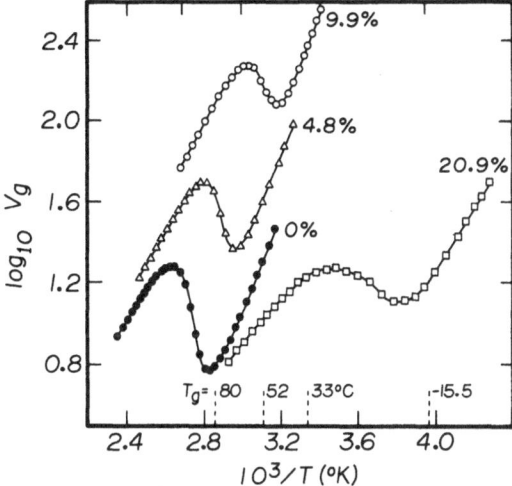

Fig. 2. Retention diagrams for ($\circ$, $\triangle$, $\bullet$) n-decane and ($\square$) n-pentane on poly(vinyl chloride) plasticized with dioctyl phthalate

diene and styrene (12). Some results on poly(vinyl chloride) stationary phases plasticized with dioctyl phthalate are shown in Fig. 2, including the transition temperatures as measured by differential scanning calorimetry (DSC). It was found that, in all cases investigated, an excellent correlation was observed between the temperature of reversal and the T_g of the stationary phase.

From the available experimental evidence it was found (14) that T_1, the temperature at which the first departure from linear behavior was observed, was in better agreement with the T_g of the stationary phase than the temperature of the minimum. T_1 as so-defined is thus the transition temperature T_g as determined by inverse gas chromatography. This finding is further supported by theoretical arguments. Departure from linearity corresponds to the first measurable contribution of bulk sorption to the total retention, indicating the onset of penetration of the solute into the bulk of the polymer. Furthermore, a large number of retention diagrams were reported (11–13) for which no minimum could be detected but exhibiting an unmistakable departure from linearity. The conditions leading to a pronounced minimum are discussed later in this review.

To establish with accuracy the temperature of first deviation from linearity, Braun and Guillet (13) computed the slopes of the retention diagram at several temperatures through T_g. It was expected that at T_1 a drastic change in slope should be observed. Figure 3 reproduces their data for a polystyrene = hexadecane system for which no minimum was recorded. Despite a most gradual departure from linearity in the retention diagram, the transition temperature T_1 was quite easily detected through the change in slope.

Transition temperatures of polyolefin stationary phases were reported by Galassi and Audisio (15) and Braun and Guillet (16). Galassi and Audisio (15) determined both melting and glass transition temperatures for isotactic, syndiotactic and atactic polypropylene. Braun and Guillet (16) reported glass transition temperatures in good agreement with accepted literature values for poly(isobutylene), poly(1-butene), polypropylene and a copolymer of ethylene and propylene. Their studies on station-

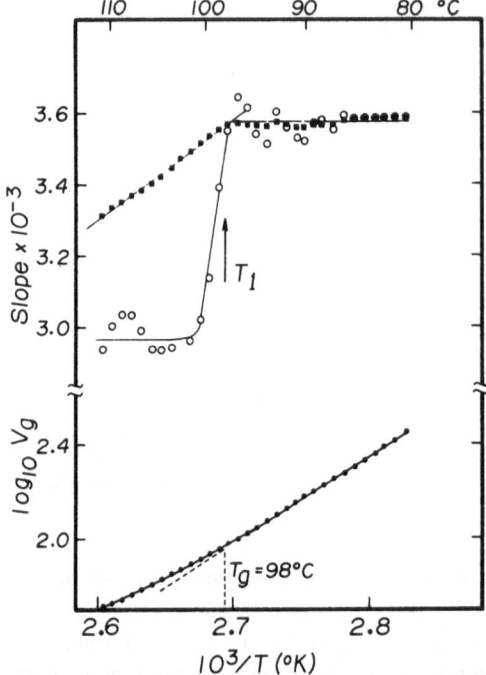

Fig. 3. (■) Overall and (○) derivative slopes of (●) retention diagram for n-hexadecane on polystyrene

ary phases of polyethylene of varying densities indicated a transition temperature in the vicinity of $-30\,°C$, increasing slightly with the crystalline content of the polymer. The corresponding retention diagrams for low and medium density polyethylenes are given in Fig. 4.

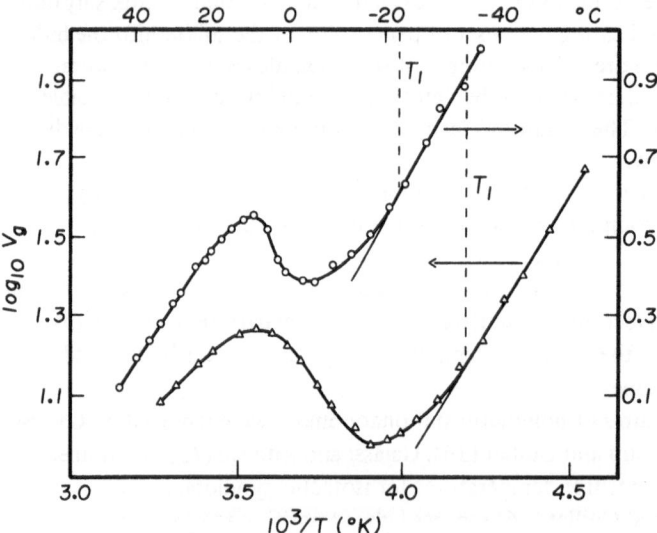

Fig. 4. Retention diagrams for (△) 2-methyl butane on low density polyethylene and for (○) n-pentane on medium density polyethylene

Liebman et al. (*17*) detected the glass transition of poly(vinyl chloride) stationary phases by recording the separation of *cis/trans* isomers as a function of temperature. It was found that no separation could be achieved in the vicinity of T_g while separation was possible both above and below T_g. Similar discontinuities in the plots of relative retention volumes were reported by Yamamoto *et al.* (*18*) for poly carbonate stationary phases of differing molecular weights.

Polymers which are available as powders of suitable size and density, may be packed directly in a column without a support for studies in the region of T_g. Hence it is not necessary to dissolve the sample prior to study. While most investigations involve thin polymer films deposited onto some inert support or the walls of a capillary tubing (*19*), studies have also been made with polymer powders (*11, 20–22*) or fibers (*21*). As a result the thermal history of the polymer sample can also be studied.

Chabert, Chauchard, Edel and collaborators (*21, 23–25*) reported three discontinuities in retention diagrams of polyester stationary phases. The transitions, denoted T_1, T_2 and T_3, were all located below the melting point of the polymer. In good agreement with results from thermal measurements, they assigned T_1 as the glass transition temperature, T_2 as a partial crystallization of the stationary phase and T_3 as the melting of metastable crystalline domains (smectic). Using powders and fibers directly as column material, they investigated the variation of the transition temperatures with the controlled heat treatment undergone prior to study. A summary of their results from inverse gas chromatography and thermal measurements is given in Table 1. It was noted that for the more highly crystalline sample detection of the transition temperatures was only possible by gas chromatography.

Other successful determinations of glass transition temperatures by inverse gas chromatography include those of Perrault *et al.* (*22*) on poly(butadiene), Nakamura *et al.* (*26*) on cellulose triacetate and Calugaru *et al.* (*27*) on poly(vinyl acetate).

While the determination of the glass transition temperature of polymer stationary phases by inverse gas chromatography seems well-established, several authors (*11, 15, 17*) reported the failure of the technique to detect known transitions. In-

Table 1. Transition temperatures of polyesters by gas chromatography (GC) and differential thermal analysis (DTA) (*24*)

	Heat treatment temperature, °C					
	Untreated	90	130	160	200	230
	Percent crystallinity					
	4.2	7	38.8	43.3	49.4	60
T_1 by GC	75–80	78–82	93	98	100	100
by DTA	78	79	87	89	Undetectable	Undetectable
T_2 by GC	119	125	130	130	130	130
by DTA	129	127	None	None	None	None
T_3 by GC	None	None	160–180	180–200	210–220	Undetectable
by DTA	None	None	142	173	217	249

stead of the expected reversal, a linear retention diagram was recorded through T_g. These findings, which could severely undermine the reliability of the method, focused attention on the importance of experimental conditions used in these experiments.

Braun and Guillet (13) investigated the effects of polymer film thickness on the retention characteristics of polystyrene-hexadecane systems. Some of their results reproducing retention diagrams covering nearly three orders of magnitude in coating thickness are given in Fig. 5. It was found that, as the coating thickness was decreased, reversal from the normal linear behavior became less pronounced, leading even to the disappearance of any minimum. At low enough surface coverage, for a loading of 0.02% polystyrene on Chromosorb G, a linear retention diagram was obtained through T_g. They further noted that T_1, the temperature of first deviation from linearity, remained constant while the temperature of the minimum varied significantly.

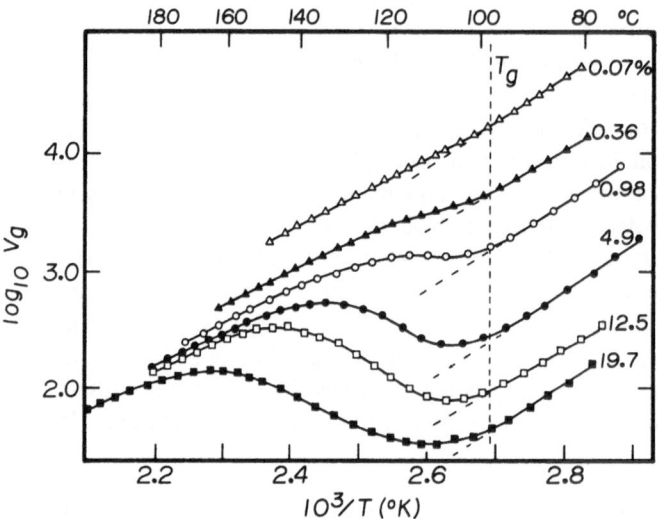

Fig. 5. Effect of coating thickness on retention diagram for *n*-hexadecane on polystyrene

The variation in the shape of the retention diagram with polymer film thickness was accounted for by a quantitative analysis of the retention data above and below T_g. It was shown that at temperatures below T_g the retention volume per unit area of the coated support was constant, irrespective of film thickness, indicating the existence of a single surface retention mechanism. At temperatures above T_g, equilibrium retention data were suitably described by the relation

$$V_N = K_b w + K_a A \tag{3}$$

where K_b and K_a are the bulk and surface partition coefficients, w and A the mass and surface area of polymer. Equation (3) was first suggested by Martin (28) to account for surface adsorption effects in GLC. As a result of a decrease in film thickness (A/w increasing), the relative contribution from bulk sorption decreases, leading to a less pronounced reversal. For extremely thin films, the second term becomes predominant, resulting in a linear retention diagram through T_g.

The importance of the solute molecule in the detection of the glass transition has been the object of several communications (1, 11, 15, 29). Simidsrød and Guillet (1) indicated that the solute must be a nonsolvent of the polymer to observe reversal from the linear behavior. It was later found that linear retention diagrams could be recorded with nonsolvents, even for thick polymer films (11, 15). Furthermore, solvents of the polymer have been used successfully to detect the glass transitions of polystyrene (30) and poly(acrylonitrile) (11). It was recently indicated (29) that the solubility of the solute in the polymer is only an incomplete characterization of the system considered. The magnitude of interaction of the probe with both bulk and surface ought to be determined, as expressed in Eq. (3). To maximize reversal, the bulk partition coefficient should be as large and the surface partition coefficient as small as possible. Braun and Guillet (29) suggested that the ratio of K_b to K_a, at T_g, should provide a satisfactory description of the chromatographic behavior near T_g. The inverse of this ratio, representing the excess concentration of solute on the surface over that in the bulk, was related (29, 31) to the limiting value of the polymer-solute solution surface tension,

$$(\partial\gamma/\partial w_1)^\infty = -(RT/M_1)(K_a/K_b) \tag{4}$$

where γ is the surface tension, w_1 and M_1 the weight fraction and molecular weight of the solute and T the temperature. It was shown (29) that the variation of the ratio K_b to K_a, computed for several polymer-solute systems (Table 2), correlated well with the observed reversal from linearity. The solubility, expressed in Table 2 as an activity coefficient, was included for the purpose of comparison. It was concluded that, in general, both solvents and nonsolvents of the polymer could be suitable for such experiments.

The origin of nonequilibrium conditions at temperatures immediately above T_g appears at present well-understood (32–35). Nonequilibrium arises when diffusion

Table 2. Bulk to surface retention ratios and limiting surface tensions for selected polymer-solute systems (29)

Polymer	Solute	T_g, °C	$(K_b/K_a)_{T_g}$	$(\partial\gamma/\partial w_1)^\infty_{T_g}$ dyne/cm	$(a_1/w_1)^\infty_{T_g} + 50$
Polystyrene	n-Hexadecane	98	27.4	− 0.50	16
Poly(N-isopropyl acrylamide)	n-Hexadecane	130	25.4	− 0.58	40
Poly(vinyl chloride)	n-Decane	80	11.0	− 1.9	42
Poly(methyl methacrylate), H	n-Dodecane	85	7.9	− 2.2	49
Poly(acrylonitrile)	Acetonitrile	110	1.3	−62	13

of the solute molecules into and out of the stationary phase is no longer fast enough, leading to a peak maximum retention volume depending on carrier gas flow rate. In the case of polymers near T_g, a decrease in temperature is accompanied by a very pronounced decrease of the diffusion coefficient of small organic penetrants. This, in turn, leads to nonequilibrium gas chromatography.

Gray and Guillet (*32*) investigated the effects on peak shape of a rapidly decreasing diffusion coefficient in a thick polymer stationary phase, assuming that the variation of the diffusion coefficient with temperature was that given by free-volume theories (*36*). It was found that as temperature was decreased the eluted peaks were no longer symmetrical, resulting in a decrease of the peak maximum retention volume from its equilibrium value, as observed experimentally. More recently, Courval and Gray (*33*) included the effects of surface adsorption in deriving peak shapes in the vicinity of T_g. It was found that the theoretical predictions were in good agreement with the experimentally-observed peak shapes. It was also indicated that for strongly interacting probes at temperatures near T_g, the contributions of surface adsorption and bulk sorption to total retention may no longer be considered independently, as done in Eq. (3).

The extent of the nonequilibrium region, segment BC of Fig. 1, is dependent on both the polymer-solute system (*34*) and experimental parameters (*13, 33*). As the thickness of the polymer film or the flow rate of carrier gas increases, the maximum of the retention diagram shifts towards higher temperatures (*13*). Equilibrium retention data can, however, be obtained by extrapolation of retention volumes to zero flow rate. Braun and Guillet (*35*) have developed a model of the chromatographic behavior of polymers near T_g which reproduces most features observed experimentally. The effects of the magnitude and temperature dependence of the diffusion

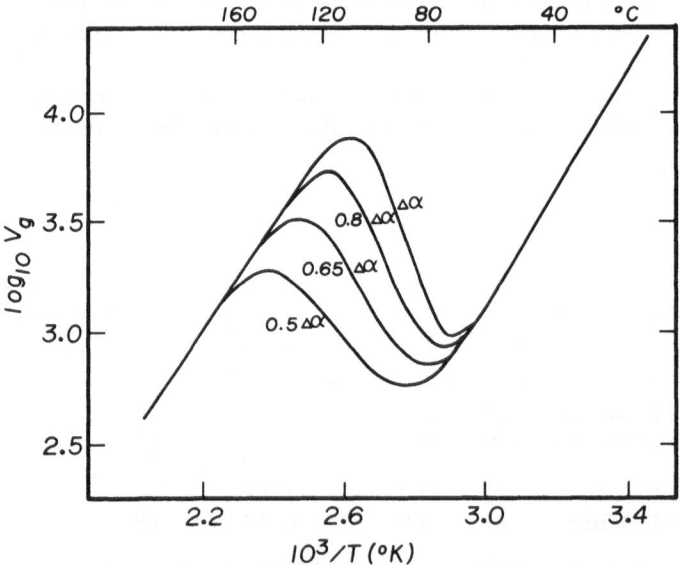

Fig. 6. Computed retention diagrams: effect of expansion coefficient, $\Delta\alpha = \alpha_l - \alpha_g$

coefficient of the solute at temperatures above T_g computed with this model are shown in Fig. 6. In the free volume treatment, the diffusion coefficient increases with $\Delta\alpha$, the difference in thermal expansion of the liquid and glassy polymer. Depending on the value of $\Delta\alpha$ considered, the extent of nonequilibrium can be drastically affected. A case in point is poly(isobutylene) for which nonequilibrium has been found to span a range of more than $120\,^\circ C$ (34). For most polymers, under normal conditions of coating thickness and carrier gas flow rate, equilibrium conditions are established at temperatures in excess of $T_g + 50^\circ$.

In summary, it may be concluded that the GC method is quite general and reliable for detecting glass, melting and possibly other transitions in polymers. Qualitative information can also be obtained about the temperature coefficient of free volume increase above T_g. The behavior is sufficiently characteristic to be suitable for distinguishing T_g from other transitions in polymers.

IV. Polymer Solution Thermodynamics

At temperatures above T_g, or T_m for a semicrystalline polymer, the magnitude of the retention volume is a direct measure of the solubility of the probe in the polymer. At infinite dilution of the solute the relation between the bulk retention volume and the activity coefficient is (5, 37, 38)

$$\ln\gamma_1^\infty = \ln(273.2R/V_g p_1^0 M_2) - (p_1^0/RT)(B_{11}-V_1) + (P_0 J_3^4/RT)(2B_{13}-V_1) \qquad (5)$$

where γ_1^∞ is the mole fraction activity coefficient, p_1^0 and V_1 the vapor pressure and molar volume of the solute, B_{11} and B_{13} the second virial coefficients, M_2 the molecular weight of the stationary phase, P_0 the outlet pressure and J_3^4 a correction factor for gas compressibility. It should be noted that, depending on actual experimental conditions, higher order terms may have to be included in Eq. (5) (37, 38). The last term of Eq. (5) is very small in most cases and will be omitted in subsequent derivations. Gas chromatography has been used extensively to provide accurate thermodynamic data on low molecular weight compounds.

In the case of polymers, the inclusion of M_2 in Eq. (5), the molecular weight of the stationary phase, creates some difficulties (39, 40). It is unclear which of the number-average, M_n, or weight-average, M_w, molecular weight ought to be used. Moreover, the variation of the activity coefficient with molecular weight as predicted by Eq. (5) is quite unrealistic. As M_2 increases to infinity (crosslinked polymer), the activity coefficient would tend towards 0. This problem was overcome in a treatment due to Patterson et al. (39). The choice of a mole fraction as composition variable becomes inappropriate when the dissimilarity in size of the component is too large. Introducing a weight fraction activity coefficient, defined as a ratio of an activity, a_1, to a weight fraction, w_1, Eq. (5) can be rewritten

$$\ln(a_1/w_1)^\infty = \ln(273.2R/V_g p_1^0 M_1) - (p_1^0/RT)(B_{11}-V_1) \qquad (6)$$

where M_1 is the molecular weight of the solute. Although the molecular weight of the polymer no longer appears explicitly, its effects would be felt through the measured retention volume. The activity coefficient is now unambiguously defined.

The link between GC quantities and the interaction parameters of solution theories is readily established (39). In statistical theories of solution thermodynamics, the solute activity is expressed as the sum of two terms, a combinatorial entropy and a noncombinatorial free energy of mixing. In the Flory-Huggins approximation one has,

$$\ln a_1 = (\ln a_1)_{\text{comb}} + (\ln a_1)_{\text{noncomb}} = [\ln \phi_1 + (1 - \tfrac{1}{r})\phi_2] + \chi \phi_2^2 \tag{7}$$

where ϕ_i is the volume fraction of component i and χ the interaction parameter. r is the ratio of molar volumes of the polydisperse polymer and the solute,

$$r = [(V_2)_n/V_1] = [(M_2)_n v_2/V_1] \tag{8}$$

where $(V_2)_n$ and v_2 are the number average molar volume and specific volume of the polymer. Combining Eqs. (6) and (7), one has, at infinite dilution ($\phi_2 \rightarrow 1$),

$$\chi = \ln(273.2 R v_2 / V_g p_1^0 V_1) - [1 - V_1/(M_2)_n v_2] - (p_1^0/RT)(B_{11} - V_1) \tag{9}$$

For high molecular weight stationary phases, the second term of Eq. (9) becomes equal to -1.

More recent polymer solution theories (41, 42) recognize the importance of the free volume dissimilarity of the solute and polymer. This effect, first introduced in theories of solution by Prigogine and collaborators (43), has important thermo-dynamic consequences. Flory and collaborators (42) now suggest that $(\ln a_1)_{\text{noncomb}}$ be composed of two terms, an equation of state and a contact interaction contribu-tion. The newly defined parameter, denoted χ^*, becomes

$$RT \chi^* = p_1^* V_1^* \left\{ \left[3 \tilde{T}_1 \ln\left(\frac{\tilde{v}_1^{1/3} - 1}{\tilde{v}_2^{1/3} - 1}\right) + \tilde{v}_1^{-1} - \tilde{v}_2^{-1} \right] + (X_{12}/p_1^* \tilde{v}_2) \right\} \tag{10}$$

with

$$\tilde{v}^{1/3} = \frac{1 + \alpha T}{3(1 + \alpha T)} \quad \text{and} \quad \tilde{T} = T/T^* = \frac{\tilde{v}^{1/3} - 1}{\tilde{v}^{4/3}} \tag{11}$$

where α is the thermal expansion coefficient. The quantities with tilde and asterisks are reduced quantities and reduction parameters, respectively. X_{12}, the contact interaction parameter, arises from potential differences between the components in solution. Using these segment fractions and "core" volumes (v^*), Eq. (9) becomes

$$\chi^* = \ln\left(\frac{273.2 R v_2^*}{V_g p_1^0 V_1^*}\right) - \left[1 - \frac{V_1^*}{(M_2)_n v_2^*}\right] - (p_1^0/RT)(B_{11} - V_1) \tag{12}$$

As is apparent from its definition, the newly defined parameter is larger than the former χ parameter. Combining data on the pure components and an experimental retention volume, the interaction parameters χ^* and X_{12} can be derived.

From the variation of the interaction parameter or the activity coefficient with temperature, the partial molar heat of mixing at infinite dilution of the probe, Δh_1^∞, can be computed

$$\Delta h_1^\infty = R \frac{\partial \chi^*}{\partial(1/T)} = R \frac{\partial \ln(a_1/w_1)^\infty}{\partial (1/T)} \tag{13}$$

It should be noted that Eq. (13) does not apply to the interaction parameter based on volume fractions (χ), due to the inclusion of a temperature dependent combinatorial entropy. In computing partial molar heats of mixing at infinite dilution, it is essential that the correction for gas phase nonideality (B_{11}) be included, owing to the magnitude of Δh_1^∞ ($\leqslant 100-300$ cal/mol).

A prime objective of early studies of polymer stationary phases was to assess the reliability of GC-derived activity coefficients and interaction parameters. Conventional static methods of measurement are difficult and time-consuming, and data on relatively few systems are available for the purpose of comparison. The facility with which such data can be obtained by gas chromatography should remedy this situation. In order to achieve a meaningful comparison, results from static methods must be extrapolated to infinite dilution, owing to the concentration dependence of the interaction parameter.

Schreiber and collaborators (44, 45) reported such comparisons of interaction parameters obtained by GC and static methods for poly(dimethyl siloxane) (44) and natural rubber (45). A summary of their results for hydrocarbon solutes in poly(dimethyl siloxane) is given in Table 3 for both χ (χ^*) and X_{12} interaction parameters.

Table 3. Flory χ (χ^*) parameters and X_{12} contact interaction parameters in poly(dimethyl siloxane) at 25 °C

Solute	χ	χ^*	χ^*_{vap}[a]	χ_{ratio}[b]	X_{12} (cal/cc)	X_{12}[c]
n-Pentane	0.409	0.513	0.51	1.096	2.88	1.1
n-Hexane	0.448	0.524		1.037	2.93	1.1
n-Heptane	0.497	0.556	0.53	1.014	2.99	1.4
n-Octane	0.556	0.600	0.57	1.005	3.35	1.6
2-Methyl butane	0.392	0.440	0.48	1.051		
Benzene	0.814	0.864	0.87	1.012	8.33	5.6
Toluene	0.802	0.833	0.82	1.006	6.91	3.5
p-Xylene	0.800	0.822	0.80	1.002	5.93	2.0
Ethyl benzene	0.828	0.833	0.80	1.002	6.03	1.8

[a] Vapor sorption measurements, Ref. (46).
[b] Ratio of corrected to uncorrected χ parameter.
[c] From Ref. (47).

These are compared with the data of Chahal *et al.* (*46*) (χ^*) and Morimoto (*47*) (X_{12}). It was found that agreement between both methods of measurement was very good for χ^* parameters. Although the magnitude of X_{12} parameters differed significantly, it was noted that the trend in both sets of data was very similar. It should be noted that the contact interaction parameters were obtained on the opposite ends of the composition range and a comparison of absolute values may not be too significant. Summers *et al.* (*44*) further commented on the importance of including the correction for vapor phase nonideality by computing the ratio of corrected to uncorrected χ parameters (Table 3). It was found that, if an accuracy of better than 5% in χ was required, this correction should be made for solute vapor pressures in excess of 200 mm Hg.

Poly(isobutylene) stationary phases have been investigated by a great many workers (*48–52*) and afford an interesting comparison. Hammers and DeLigny (*48, 49*) analyzed their results in terms of the modified Prigogine (*41, 43*) and Flory (*42*) theories. It was found that for both linear and branched alkanes a two parameter Prigogine model accounted suitably for the experimental data while the Flory theory gave unrealistic results. Marcille *et al.* (*50*) investigated the variation of the interaction parameter with the molecular weight of the stationary phase, observing an increase in χ as the molecular weight increased from 500 to 2500. Table 4 summarizes the interaction parameters for several hydrocarbon solutes obtained for high molecular weight samples by several groups of workers (*48, 51, 52*). The results of Eichinger and Flory (*42*) were included for the purpose of comparison. While there are a few minor discrepancies, the agreement observed between GC-derived interaction parameters and static results is most satisfactory.

Table 4. Flory χ^* Parameters in poly(isobutylene) (*52*)

Solute	χ^*			
	Data of Leung and Eichinger (*52*)	Data of Hammers and DeLigny (*48*)	Data of Newman and Prausnitz (*51*)	Vapor sorption measurements (*42*)
n-Pentane	0.90	0.85	0.87	0.88
n-Hexane	0.79	0.72		
n-Heptane	0.69	0.69		
n-Octane	0.63	0.67		0.65
n-Nonane	0.59	0.64		
Benzene	0.99	1.00	1.03	1.13
Cyclohexane	0.67	0.55	0.55	0.52

Schreiber, Tewari and Patterson (*53*) reported interaction parameters of more than 20 hydrocarbons in linear and branched polyethylenes, at temperatures above the melting point. The corresponding χ parameters are given in Table 5. Despite the chemical identity of the components, substantial interaction parameters were obtained. This was attributed to the large contribution arising from the free volume dissimilarities of the components. Indeed it proved possible to correlate the magnitude

Table 5. Interaction parameters for linear (LPE) and branched (BPE) polyethylenes (53)

Solute	χ					
	LPE			BPE		
	145.4 °C	152.6 °C	Avg. $\bar{\chi}$	120.0 °C	145.1 °C	Avg. $\bar{\chi}$
3-Methylhexane	0.421	0.385	0.403	0.328	0.335	0.332
n-Octane	0.366	0.350	0.358	0.307	0.300	0.304
2-Methylheptane	0.393	0.391	0.392	0.331	0.340	0.336
3-Methylheptane	0.373	0.364	0.369	0.308	0.300	0.304
2,4-Dimethylhexane	0.391	0.360	0.376	0.336	0.323	0.330
2,5-Dimethylhexane	0.425	0.379	0.402	0.354	0.352	0.353
3,4-Dimethylhexane	0.319	0.296	0.308	0.247	0.249	0.248
2,2,4-Trimethylpentane	0.411	0.392	0.402	0.341	0.336	0.339
n-Nonane	0.347	0.330	0.339	0.275	0.282	0.279
2,2,4-Trimethylhexane	0.368	0.327	0.348	0.288	0.284	0.286
n-Decane	0.317	0.306	0.312	0.254	0.255	0.255
n-Dodecane	0.293	0.277	0.285	0.227	0.240	0.234
Toluene	0.393	0.395	0.394	0.336	0.338	0.337
Ethylbenzene	0.368	0.372	0.370	0.326	0.328	0.327
p-Xylene	0.319	0.322	0.320	0.270	0.277	0.274
m-Xylene	0.340	0.340	0.340	0.288	0.294	0.291
Mesitylene	0.290	0.274	0.282	0.244	0.249	0.247
Tetralin	0.325	0.318	0.322	0.289	0.279	0.284
cis-Decalin	0.080	0.063	0.072	0.028	0.027	0.027
trans-Decalin	0.061	0.045	0.053	0.017	0.003	0.010

of the measured interaction parameter with the difference in thermal expansion coefficients of the solutes and polymer. In agreement with an earlier study by these authors (54) on alkane systems, it was concluded that the methyl-methylene interchange energy, hence differences between linear and branched solutes, should be small. It was noted (54), however, that other effects, such as correlations of molecular orientations, could differentiate between linear and branched alkanes. Such effects, which could be detected for heats and entropies of mixing, would, however, cancel out in free energies and hence χ's.

Lichtenthaler et al. (55) determined interaction parameters for 22 solutes in poly(dimethyl siloxane) to test several expressions of the combinatorial entropy of mixing [Eq. (7)]. The magnitude of the interaction parameter is indeed directly dependent on the evaluation of the combinatorial contribution. The combinatorial contribution was computed following both the Flory-Huggins approximation and the multiple-connected-site model recently developed by Lichtenthaler, Abrams and Prausnitz (56). This model, which retains the Flory-Huggins term, also corrects for the bulkiness of the components of the mixture. Interaction parameters were computed through both approximations, showing the sensitivity of the results to the model chosen.

Hammers and DeLigny (57, 58) reported their studies on the determination of partial molar heats of mixing by gas chromatography and their application to test polymer solution theories. There is as yet little agreement in the technical literature as to whether GC can provide partial molar heats of mixing with any degree of accuracy. To gain further insight into this problem, the partial molar free energies of evaporation of the solute from the pure liquid and from the polymer solution were separately fitted to the following Taylor series expansion,

$$\Delta G/T = \Delta G_\theta / \theta - \Delta H_\theta \left(\frac{1}{\theta} - \frac{1}{T}\right) + \Delta C_{p\theta} \left[1 - (\theta/T) - \ln (T/\theta)\right] +$$

$$\frac{1}{2} (d\Delta C_p/dT)_\theta [(\theta^2/T) - T + 2\theta \ln (T/\theta)] \qquad (14)$$

where $\Delta G, \Delta H$ and ΔC_p have their usual thermodynamic significance, θ being some reference temperature. Three cases were considered, a zero, constant and constantly varying heat capacity. To detect with accuracy any temperature dependence, retention data were collected over a range of 150°. It was found that the experimental data could not be accommodated with a zero heat capacity. On the other hand, the temperature coefficient of the heat capacity, while nonzero, could not be determined with enough accuracy. As a result, data for several hydrocarbons in poly(dimethyl siloxane) were analyzed with a constant heat capacity. The so-determined partial molar heats of mixing were compared with results from static methods (Table 6). While the error on the GC value is much larger, the absolute values compare favorably. It is expected that, under appropriate conditions, GC could provide accurate heats of mixing, depending of course on the absolute value of the quantities involved.

As a result of the growing interest in the GC route to obtain information on polymer-solute systems, a large body of data, activity coefficients and/or interaction parameters, has been reported. Polystyrene (51, 59), poly(vinyl chloride) (60), polyethylene (60–62), poly(ethylene oxide) (63) and copolymers of ethylene with propylene and vinyl acetate (62) have been studied with a variety of probes.

Table 6. Partial molar heats of mixing (Δh_1^{∞}) of some alkanes in poly(dimethyl siloxane) at 25 °C (57)

Solute	$\Delta h_1^{\infty}/RT$	
	Calorimetry (46)	GC
n-Pentane	−0.02 ± 0.02	0.02 ± 0.09
n-Hexane	0.18 + 0.02	0.23 + 0.06
n-Heptane	0.28 + 0.02	0.23 + 0.06
n-Octane		0.29 + 0.12
2,2,4-Trimethylpentane		0.44 + 0.07
n-Decane	0.80 + 0.05	0.49

Despite the good agreement usually observed between GC and static results, attention has recently been drawn to discrepancies in GC derived interaction parameters. Lichtenthaler et al. (64) reported markedly higher retention volumes on poly(dimethyl siloxane), hence lower χ parameters, than did Summers et al. (44). An ensuing interlaboratory comparison (65) of retention data did not entirely resolve this difficulty, indicating that column packing procedure appeared as a crucial step. Lichtenthaler, Liu and Prausnitz (66) have also expressed concern about possible orientation effects of the inert support. The conditions prevailing in a rather thin polymer film spread on an inert support may thus be different from those in a thick polymer slab. In order to increase the polymer film thickness it was suggested (66) that capillary columns be used. In way of further commentary it may be added that the determination of reproducible interaction parameters calls for an accurate determination of the total mass of polymer in the column. The most common methods, extraction or ignition of the coated support, require a blank conducted in similar conditions. Failure to correct for this effect could bring about a sizable error. Other sources of error include flow rate dependent retention volumes and surface adsorption effects. It should be noted in this respect that the coverage ratio (grams polymer/grams support) is a poor criterion for gauging the likelihood of surface adsorption. It is well known that common commercial supports vary in specific surface area, from about 0.5 m^2 g^{-1} to as much as 10 m^2 g^{-1} and a twenty-fold change in film thickness can be obtained at identical coverage ratio. Retention volumes should be obtained on columns of differing surface to volume ratios.

In addition to theoretical studies, GC has been used for more practical applications. Reichert (67) and Newman and Prausnitz (68) have studied the interactions in paint films by gas chromatography. The conditions prevailing in a GC experiment should approximate very closely the drying of paint films. Table 7 summarizes some of the results of Newman and Prausnitz (68) on several technologically important polymers. The solutes were solvents commonly used in the paint industry.

McCathern and Thompson (69) studied the swelling of vulcanizates in various media in relation to the GC determined interaction parameters. It was found that for over 50 polymer-solute pairs a most satisfactory correlation was observed between the volume increase and the χ parameter. Considering the concentration range covered (volume increases of up to 400%) and the concentration dependence of the

Table 7. Weight fraction activity coefficients for selected polymer-solute systems at 150 °C $(a_1/w_1)^\infty$ (68)

Solute	Polymer						
	Poly(methyl methacrylate)	Alkyd resin, Beckosol®-23	Poly(n-butyl methacrylate)	Epoxy resin, Eponol®-55	Polyamide, Epotuf®	Poly(vinyl acetate)	Polyurethane
n-Hexane		22.0	10.4				
Cyclohexane		14.1					
Benzene	8.46	6.56	4.59	7.81	6.13	5.86	7.41
Toluene	8.99	6.74	4.54	8.05	6.29	6.42	7.59
p-Xylene	10.15		4.70				
Chloroform	2.87	2.99	1.93	3.96	2.12	2.16	
1,2-Dichloroethane	4.21	4.11	3.28	4.09	3.01		
Isopropylalcohol		9.01	9.67	13.5	5.88	7.36	
n-Butylalcohol		7.48	7.67	9.88	4.38	6.56	9.57
Methyl ethylketone		6.52	7.44	8.62	8.32	6.34	
Methyl isobutyl ketone		7.76	7.45	10.3	9.17	8.39	11.46
Cyclohexanone		3.74	5.08	4.0		4.63	4.77
Methyl cellosolve		6.75	10.7	8.40	4.77	6.20	8.19
Cellosolve solvent		6.46	8.69	7.88	4.66	6.04	
Tetrahydrofuran							
p-Dioxane			4.64		4.70		5.52
Vinyl acetate						5.71	4.71
Ethyl acetate		6.70	6.66	8.02	8.43	5.92	
n-Butyl acetate		6.72	5.87	8.22	8.47	6.88	9.03
N,N-Dimethyl formamide		3.08		3.87			
Acetonitrile		11.7	17.3	14.8	13.7		4.23

χ parameter, a perfect correlation was not to be expected. From the correlation curve so-established the swelling of experimental vulcanizates could be predicted with reasonable accuracy from the easily determined χ parameter. The GC method should be particularly valuable for testing experimental samples, whenever only small amounts are available. For non-crosslinked materials, the magnitude of the χ parameter is a direct indication of the solubility of the polymer in any given solvent. The dividing line between solvents and non-solvents of the polymer can be drawn at approximately $\chi = 0.5$, the smaller the values of χ below this limit, the better the solvent. It is a simple matter to measure the retention characteristics of a series of probes and thus determine a suitable solvent for any polymer. It should be noted, however, that the temperature at which the GC determination is possible ($T > T_g + 50$) may sometimes exceed the temperature of interest and the dependence of χ on temperature may have to be assessed.

The experimental work reported up to this point has dealt exclusively with solutes at infinite dilution in the polymer stationary phase. Recent theoretical advances (38, 70) have made it possible to extend the range of gas chromatography to finite concentration of the solute in the stationary phase. In the case of polymers such studies should provide a direct insight into the concentration dependence of the interaction parameter and allow for comparison with static results without recourse to extrapolations to infinite dilution.

Following the theoretical treatment of Conder and Purnell (38, 70) of gas chromatography at finite concentration, Brockmeier et al. (71, 72) applied the method to polymer stationary phases. For the technique of elution on a plateau of finite concentration, the weight fraction activity coefficient is given by the relation (71),

$$a_1/w_1 = P \psi/p_1^0 w_1 \tag{15}$$

where ψ is the mole fraction of solute above the stationary phase and P the mean pressure in the column. The weight fraction of the solute in the stationary phase (w_1) is obtained from the experimentally determined solubility isotherm. For the χ parameter one has, at finite concentration,

$$\ln(a_1/w_1) = \ln[(1 - w_1) r + w_1] + \frac{1 - w_1}{(1 - w_1) + w_1/r} + \chi\left(\frac{1 - w_1}{(1 - w_1) + w_1/r}\right) \tag{16}$$

where r is the ratio of the molar volume of the components. The relationship between Eqs. (16) and (7) is quite direct. The concentration dependence of χ is not explicitly considered in Eq. (16). Brockmeier et al. (71, 72) investigated both amorphous and semicrystalline polymers over a range of solute concentration. Comparison with other GC work, in the infinite dilution limit, was very good for the systems investigated. Some of their results on the sorption of n-decane by polyethylene are shown in Fig. 7. It was found that, in good agreement with other studies, the interaction parameter had little concentration dependence.

Finite concentration GC offers additional advantages. The temperature range amenable to experiment is extended, as the limitations set by nonequilibrium chromatography are somewhat relaxed. In the presence of a finite amount of solute,

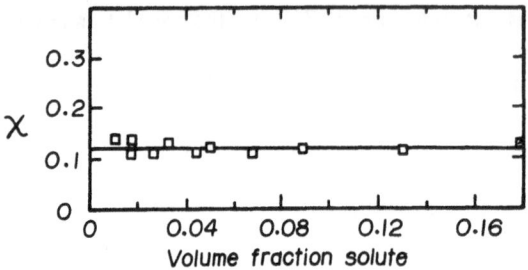

Fig. 7. Interaction parameter as a function of composition for n-decane in high density polyethylene

plasticization of the polymer takes place, depressing the glass transition temperature of the system. As a result equilibrium conditions can be sustained down to lower temperatures. In addition, the prime experimental data are used to compute the solubility isotherm.

More recently Chang and Bonner (73) reported their results on the sorption by poly(ethylene oxide) stationary phases at finite concentration of solute. Earlier work by Bonner and collaborators (63) at infinite dilution had indicated a lack of reproducibility for retention volumes measured on poly(ethylene oxide). The results were found to vary with the age of the columns and this was attributed to losses of polymer from the column. To correct for the weight loss, retention volumes for a reference solute were determined at regular intervals, under identical conditions, and compared with their initial value (new column), enabling a computation of the weight loss. A summary of their results for the sorption of benzene at 70 °C is given in Table 8 for solute weight fractions up to 0.39. The infinite dilution activity coefficient and corresponding weight loss of polymer were included for both runs. It was found that good agreement was obtained with results from static experiments, except for the highest concentrations of solute.

The application of gas chromatography to the study of polymer-polymer interactions has been recently described by Patterson, Schreiber and collaborators (74–76).

Table 8. Weight fraction (w_1) and activity coefficient (a_1/w_1) of benzene in solutions of poly(ethylene oxide) (PEO) (73)

70.0 °C First run		70.0 °C Second run	
w_1	a_1/w_1	w_1	a_1/w_1
0.06163	4.3118	0.05005	4.1031
0.06711	4.2311	0.08908	3.7426
0.0991	3.8095	0.1422	3.3435
0.1387	3.4739	0.2006	2.9955
0.1926	3.0925	0.2649	2.7017
0.261	2.6898		
0.3881	2.1527		

$(a_1/w_1)^\infty = 4.572$ $(a_1/w_1)^\infty = 4.572$

Conditions: 24% PEO wt loss   Conditions: 9.5% PEO wt loss

In such ternary systems comprising a solute and a mixed stationary phase (polymers 2 and 3), the polymer-polymer interaction is derived through the interactions of the solute with the mixed and the pure stationary phases. Note that in ternary systems index 3, formerly applied to the carrier gas [Eq. (5)], is now given to a polymer.

At infinite dilution ($\phi_1 \rightarrow 0$), the activity of the solute in a mixed stationary phase becomes (74)

$$\ln(a_1/\phi_1)^\infty = \{[1 - (r_1/r_2)] \phi_2 + [1 - (r_1/r_3)] \phi_3\} \text{ comb term} +$$

$$[\chi_{12}\phi_2 + \chi_{13}\phi_3 - r_1(\chi_{23}/r_2)\phi_2\phi_3] \equiv \chi_{1(23)} \text{ noncomb term} \qquad (17)$$

where χ_{ij} is the Flory parameter and r_i the number of segments in component i. When components 2 and 3 are polymers, χ_{23} becomes inconvenient, increasing with the molar volume V_2. It is advantageous to normalize the interaction to the unit volume of a component by introducing the ratio

$$\alpha_{ij} = \chi_{ij}/V_i \qquad (18)$$

α_{ij} should be symmetrical and dependent only on the nature of i and j, irrespective of the chain length. One then has for the interaction parameter [cf. Eqs. (6) to (9)],

$$\chi_{1(23)} = \ln \frac{273.2R(w_2 v_2 + w_3 v_3)}{p_1^0 V_g V_1} - \left(1 - \frac{V_1}{V_2}\right)\phi_2 - \left(1 - \frac{V_1}{V_3}\right)\phi_3 -$$

$$- (p_1^0/RT)(B_{11} - V_1) = [(\chi_{12}/V_1)\phi_2 + (\chi_{13}/V_1)\phi_3 - (\chi_{23}/V_2)\phi_2\phi_3]V_1 \qquad (19)$$

where v_i is the specific and V_i the molar volume of component i. χ_{12} and χ_{13} are determined from the corresponding pure stationary phases, allowing for the determination of χ_{23}, the polymer-polymer interaction parameter, from a measurement on the mixed stationary phase. To yield meaningful polymer-polymer interaction parameters it is essential that the polymers be compatible.

As previously described for binary systems, volume fractions may be replaced by segment fractions which deal with the hard core volume of the components. A new interaction parameter, $\chi^*_{1(23)}$, and the contact interaction parameter, X_{23}, are thus introduced which can be determined by GC [cf. Eqs. (10) to (12)] from the pure (X_{13}, X_{12}) and mixed stationary phases. The equivalent of the normalized interaction parameter α_{ij} of Eq. (18) is X_{ij}/s_i where s_i is the molecular surface to volume ratio.

Table 9 summarizes some of the results of Deshpande et al. (74) on mixed tetracosane-dioctyl phthalate and tetracosane-poly(dimethyl siloxane) stationary phases. The interaction parameters so-determined afford an interesting test. According to theory both χ_{23}/V_2 and X_{23}/s_2 should be independent of the chain length for interactions between similar compounds. This was generally observed with alkanes for the contact interaction term, X_{23}/s_2, with the exception of pentane. As already mentioned the χ parameter also reflects the free volume dissimilarity of the components and χ_{23}/V_2 varies slightly with the solute considered. Interestingly enough, the magnitude of X_{23}/s_2 for tetracosane-dioctyl phthalate is nearly identical to X_{12}/s_1 for the pure dioctyl phthalate with both linear and branched alkane solutes.

Table 9. Interaction parameters for mixed stationary phases of n-tetracosane (n-C_{24}), dioctyl phthalate (DOP) and poly(dimethyl siloxane) (PDMS) (74)

| Solute | Interaction between solute (component 1) and pure stationary phase (component 2) | | | | | | | | Interaction between two components (2 and 3) in the stationary phase | | | |
| | n-C_{24} | | | | DOP | | PDMS | | n-C_{24}-DOP | | n-C_{24}-PDMS | |
	X_{12} 60°	X_{12}/s_1 ×10⁸ J cm⁻² 60°	X_{12} 75°	X_{12}/s_1 ×10⁸ J cm⁻² 75°	X_{12} 75°	X_{12}/s_1 ×10⁸ J cm⁻² 75°	X_{12} 60°	X_{12}/s_1 ×10⁸ J cm⁻² 60°	$V_1 X_{23}/V_2$ cm⁻³ 75°	X_{23}/s_2 ×10⁸ J cm⁻² 75°	$V_1 X_{23}/V_2$ cm⁻³ 60°	X_{23}/s_2 ×10⁸ J cm⁻² 60°
n-Pentane	0.32	4.5	0.32	4.4	0.76	21.2	0.45	12.5	0.86	33.9	1.01	34.6
n-Hexane	0.24	4.6	0.24	5.0	0.67	19.2	0.43	13.4	0.72	25.5	0.48	11.7
n-Heptane	0.20	4.2	0.20	4.6	0.67	18.8	0.45	13.8	0.77	25.5	0.55	12.1
n-Octane	0.17	3.4	0.17	3.8	0.68	17.6	0.49	13.4	0.87	26.4	0.64	12.6
2-Methylpentane	0.26	5.0	0.27	5.9	0.69	20.1	0.42	13.0	0.72	26.8	0.57	15.9
3-Methylpentane	0.23	4.6	0.24	5.4	0.66	19.7	0.41	13.4	0.74	27.6	0.49	13.0
2,4-Dimethylpentane	0.26	5.0	0.26	5.4	0.70	21.3	0.42	13.4	0.77	28.5	0.49	13.0
Cyclohexane	0.17	5.4	0.17	5.9	0.48	19.2	0.44	19.2	0.62	28.5	0.42	10.9
Carbon tetrachloride	0.26	10.5	0.26	10.9	0.19	6.3	0.42	20.1	0.48	22.2		
Benzene	0.51	23.4	0.48	23.0	0.16	5.4	0.62	31.4	0.43	19.2	0.37	14.6
Toluene	0.35	15.5	0.36	16.3								

Similar observations can be made for the magnitude of normalized interaction parameters between tetracosane and poly(dimethyl siloxane) as compared to the interaction of alkane solutes with pure poly(dimethyl siloxane). The consistency of these results indicates that GC should prove most valuable for the study of interactions between polymers.

More recently Patterson *et al.* (75, 76) applied the GC method to the study of interactions in the technologically important poly(vinyl chloride)-dioctyl phthalate system. By using several columns the whole composition range could be investigated. It was found that, at low plasticizer contents ($\phi \leqslant 0.25$), the normalized interaction parameter was strongly negative, indicating a high compatibility. At higher concentrations of plasticizer it became less negative, turning positive at 0.55 volume fraction, suggesting a lower compatibility limit. It was noted that the normalized interaction parameter varied significantly with the chain length of the normal alkane solutes used. As a possible explanation for this phenomenon it was suggested that either a nonrandom mixing of the components of the stationary phase and/or a preferential solution of the solute in one of the components was to be suspected.

Nesterov and Lipatov studied the compatibility of mixtures of crystallizable polymers (77) and the effects of quartz fillers on polymer-solute interactions (78). Information on the compatibility of these systems was obtained via the determination of melting points and crystallinities of the mixed stationary phases. Depending on the polymers considered a single melting transition at an intermediate temperature or distinct melting transitions for each polymer could be detected.

Olabisi (79) evaluated the miscibility of polymer pairs by gas chromatography with mixed poly(ϵ-caprolactone)-poly(vinyl chloride) stationary phases. Some interaction parameters for the pure and mixed stationary phases are given in Table 10.

Table 10. Interaction parameters for mixed stationary phases of poly(vinyl chloride) (PVC) and poly(ϵ-caprolactone) (PCL) at 120 °C (79)

Solute	PCL		PVC		PCL–PVC (50:50)		
	$\overset{*}{\chi}_{12}$	X_{12} cal/cm^3	$\overset{*}{\chi}_{12}$	X_{12} cal/cm^3	X_{23}	$\overset{*}{\chi}_{23}$	X_{23} cal/cm^3
Ethanol	1.15	21.4	2.35	40.6	0.21	−0.13	−2.8
Chloroform	−0.20	−4.20	1.38	16.6	0.33	−0.09	−2.4
Methyl ethylketone	0.533	2.75	1.00	4.41	−0.10	−0.61	−6.4
Pyridine	0.175	0.239	0.939	7.34	−0.17	−0.47	−5.4
Acetonitrile	1.11	19.3	1.85	29.3	−0.40	−0.98	−9.3
Fluorobenzene	0.127	−1.61	1.25	8.82	0.24	−0.15	−2.9
Carbon tetrachloride	0.391	1.06	1.49	10.2	1.07	0.63	3.0
Hexane	1.24	7.89	1.76	8.6	1.16	0.60	2.8

The starred interaction parameters have their usual significance. It was noted that, in this instance, χ_{23}^{*} may represent a better picture of polymer-polymer interaction than χ_{23} since binary mixtures of poly(ϵ-caprolactone) and poly(vinyl chloride) are known to be stable and a small or even negative interaction parameter would be expected. Variation of the polymer-polymer interaction parameter with the solute molecule was tentatively ascribed to a nonrandom solution of the probe in the mixture.

Recently Purnell and collaborators (80) re-examined most of the results published on complexing studies of miscible low molecular weight substances. It was found that the activity coefficient of the solute in the mixed stationary phase could be described by the following relation:

$$[1/\gamma_{1(23)}^{\infty}] = [x_2/\gamma_{1(2)}^{\infty}] + [x_3/\gamma_{1(3)}^{\infty}] \tag{20}$$

where x_2 and x_3 are the mole fractions of components 2 and 3, γ_1^{∞} being the mole fraction activity coefficient [Eq. (5)]. It was shown that this relation was satisfied over the whole composition range $(0 < x_2 < 1)$ for most systems investigated. The activity coefficient of the solute in a mixed stationary phase can thus be predicted from the activity coefficients of the pure components. To account for this unexpected finding it was suggested that mixing could not be random and that the local concentration of, for example, component 2 in the 2–3 mixture was always that corresponding to pure 2, leading to a microscopic partition theory of solutions. This microscopic partition theory was further tested with solid-solid mixtures which were known to be inhomogeneous on the microscopic level. Again, the chromatographic behavior of the mixtures could be predicted from the properties of the pure components, as expressed in Eq. (20). It is to be anticipated that this finding will generate a wide interest, both from the theoretical and experimental standpoint.

V. Crystallinity of Polymer Stationary Phases

The determination of the crystallinity of polymer stationary phases is based on the differential solubility of the solute in crystalline and amorphous domains. In effect the solute "senses" only the amorphous regions, leading to an increase in retention volume with decreasing crystallinity.

Alishoev et al. (6) first observed drastic changes in the retention behavior of polyethylene and polypropylene powders in the vicinity of their melting points. Later Guillet and Stein (7) showed that retention diagrams through the melting transitions of semicrystalline polymers (Fig. 8) could be analyzed quantitatively to yield the crystallinity and melting curve of the polymer. It was found that both above and below the melting point (T_m) retention proceeded by bulk sorption, thus allowing for a comparison of retention volumes. Above T_m the polymer is completely amorphous and a linear retention diagram is obtained. By extrapolating this straight line to lower temperatures the retention volume for the theoretically amorphous polymer can be computed. Comparison with the experimental retention volume yields the amorphous fraction of the stationary phase, X_a,

$$X_a = V_g/V_g' \tag{21}$$

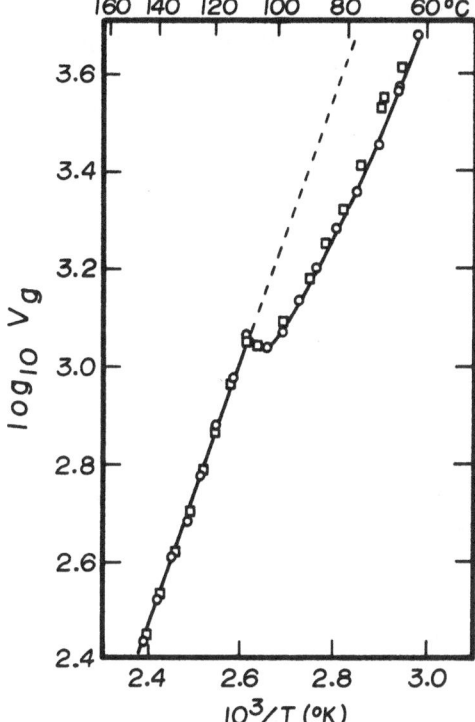

Fig. 8. Retention diagrams for n-dodecane on low density polyethylene

where V_g and V_g' are the experimental and extrapolated retention volumes, at the same temperature. For the crystallinity one has

$$\text{wt-\% crystallinity} = 100 \left[1 - (V_g/V_g') \right] \qquad (22)$$

The melting curve is obtained by determining the crystallinity at several temperatures.

The GC route is particularly attractive for it requires no a priori information on the polymer. With the exception of X-ray measurements, most methods of measurement involve a comparison of some property of the polymer, such as density, with that of the totally amorphous or crystalline material. Furthermore neither the mass of polymer in the column nor the flow rate of carrier gas need to be measured since a ratio of retention volumes is computed in Eq. (21). It should be added, however, that for the successful application of the method it is essential that the measured retention volumes correspond effectively to equilibrium bulk sorption, both above and below T_m. Low molecular weight compounds are known to exhibit apparently similar discontinuities in retention diagrams at their melting points but this is to be ascribed to a change in retention mechanism, from surface adsorption for the solid to bulk sorption for the liquid stationary phase. For a detailed discussion of retention characteristics of low molecular weight substances near their transition temperatures the reader is referred to a recent review by McCrea (81).

Comparison of crystallinities determined by gas chromatography and other methods of measurements is usually quite good. In the case of polyolefins for which density provides a well-established correlation with crystallinity, Guillet and collaborators (*7, 82–84*) demonstrated the value of the method. Crystallinities, temperatures of disappearance of the last traces of crystallinity, and melting curves for low and high density polyethylene as well as polypropylene stationary phases were in excellent agreement with results from density and thermal measurements.

The success of the method prompted the design of an "Automatic Molecular Probe apparatus" for the collection of retention data (*82*). At a preset cycle time a mixture of solute and noninteracting marker was injected into the carrier gas and the output of the GC detector was fed to an electronic peak detector. The temperature in the oven was programmed linearly and recorded with a thermocouple. Upon conversion into digital form, a printout of net retention time and temperature was obtained. Retention diagrams have been obtained with this apparatus for high and low density polyethylene (*82, 83*).

Braun and Guillet (*84*) investigated copolymers of ethylene with vinyl acetate and carbon monoxide as well as modified polyethylene waxes. While the crystallinity of such polymers can no longer be derived from their density, it was found that the GC method was equally successful with copolymers or modified polymers. Figure 9

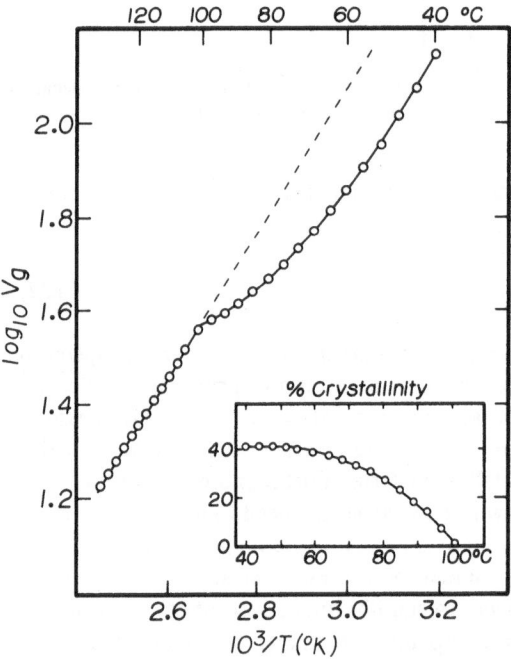

Fig. 9. Melting curve and retention diagram for *n*-heptane on poly-(ethylene-CO)

shows the retention diagram and melting curve obtained for a copolymer of ethylene and carbon monoxide. Results were in good agreement with literature data for all polymers investigated. It was also shown that polymer powders could be dispersed on an inert support and used directly in GC columns, thereby alleviating the need for dissolution of the sample prior to study. It was noted that in these studies the particle size of the polymer powder was of critical importance. At very large particle sizes ($\phi \geqslant 200 \ \mu$m) nonequilibrium conditions prevailed, even at temperatures close to melting, necessitating an extrapolation of retention volumes to zero flow rate. The method should be of value for polymer samples which have not reached their equilibrium crystallinities, and for studies of the effects of thermal history on crystallinity.

Courval and Gray (85) studied poly(ethylene oxide) stationary phases of differing film thicknesses with polar solutes. It was found that the resulting crystallinity depended on the film thickness, decreasing with decreasing film thickness. However, under these experimental conditions, surface adsorption becomes of importance and Eq. (21) is no longer applicable. As the thickness of the stationary phase is decreased, the contribution from surface adsorption increases but its relative magnitude is larger below T_m, as illustrated in Fig. 10. The apparent crystallinities computed from these

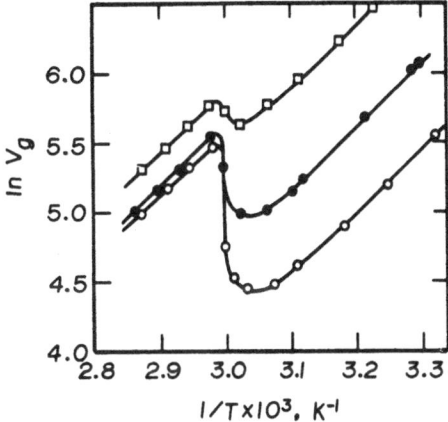

Fig. 10. Retention diagrams for *n*-propyl alcohol on poly(ethylene oxide) for (□) 0.4%, (●) 2.5%, and (○) 7.5% loadings

curves varied from 40 to 80% with surface coverage. In order to derive a meaningful value, retention data above and below T_m were extrapolated to infinite thickness of the stationary phase to obtain the actual bulk retention volume, yielding a crystallinity of 85%. A similar decrease in apparent crystallinity with decreasing film thickness was reported by Nesterov and Lipatov (86) for poly(ethylene glycol) and poly(ethylene adipate) stationary phases, but correction was not made for the surface adsorption contribution in this case.

Recently Braun and Guillet (87) evaluated the approximations involved in the determination of crystallinity by gas chromatography. While retention diagrams obtained above T_m are usually linear, the heat of vaporization of the solute, related to the slope of the retention diagrams, is known to decrease with temperature, thereby imparting a certain curvilinearity to these plots. It was shown that the effects of inexact linearity could significantly reduce the computed crystallinity if the retention diagram spanned a large temperature range. By carefully selecting experimental conditions, these effects can, however, be minimized. Furthermore an exact treatment is possible if thermodynamic information is available for the probe molecule used.

Stein, Gray and Guillet (82, 83) demonstrated the suitability of the GC method to study crystallization kinetics, through the variation of the retention volume with time. When the polymer stationary phase is cooled from the melt to a temperature below T_m, the retention volume decreases with time at the rate at which crystalline domains are being formed. The maximum possible crystallinity at a given temperature is obtained from the relation

$$(\% \text{ crystallinity})_{max} = 100\,[1 - (V_g^e/V_g')] \tag{23}$$

where V_g^e is the retention volume of the polymer once equilibrium between crystalline and amorphous phases has been reached. The percent crystallization at any time t is then

$$\% \text{ crystallization} = 100\,(V_g' - V_g^t/V_g' - V_g^e) \tag{24}$$

where V_g^t is the retention volume measured at time t. Figure 11 illustrates such results on a high density polyethylene stationary phase at several temperatures, obtained with the automatic molecular probe instrument described earlier. Good agreement was obtained between GC and published dilatometric data (82).

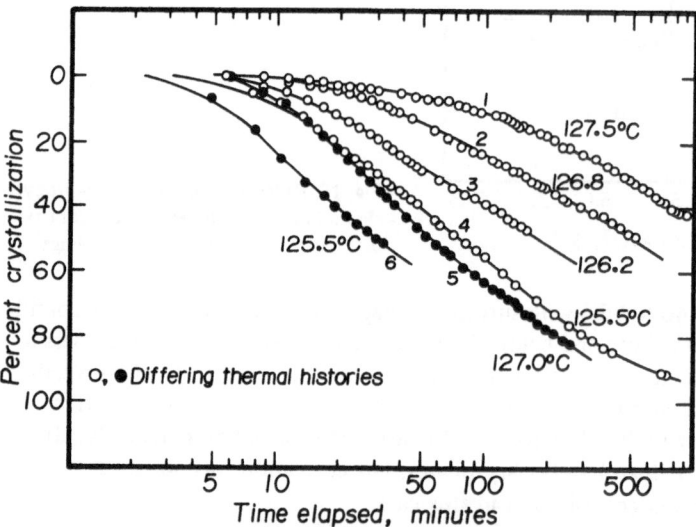

Fig. 11. Isothermal crystallization kinetics for high density polyethylene

VI. Surface Properties

The introduction of high surface area, crosslinked polymers as standard stationary phases has generated a constant interest in the study of polymer surfaces by gas chromatography (*88, 89*). At temperatures below the glass transition of the polymers where retention proceeds exclusively by surface adsorption, heats of adsorption, surface areas and adsorption isotherms can be obtained from the magnitude of the retention volume and the shape of the eluted solute peak.

Following the technique of Gluekauf (*90*), Edel and Chabert (*91*), and Gray and Guillet (*92*) determined adsorption isotherms for polymer substrates from a single unsymmetrical peak. Intuitively one would expect the peak shape and retention volume to be dependent on the adsorption isotherm, linear isotherms yielding symmetrical peaks and retention volumes independent of sample size. It can be shown that for each gas phase concentration c of eluted solute there exists a corresponding retention volume V_c given by the relation

$$a = (1/m) \int_0^c V_c dc \tag{25}$$

where a is the amount of solute adsorbed on a mass m of polymer. One side of the eluting peak should be vertical while V_c can be obtained from the other side of the peak. The isotherm relating a and c is thus determined through Eq. (25). The shape of experimental elution profiles recorded (*92*) for decane on poly(methyl methacrylate) for several sample sizes is shown in Fig. 12. It is to be noted that all front profiles fall on a common line, the rear of the peak being sharp. Figure 13

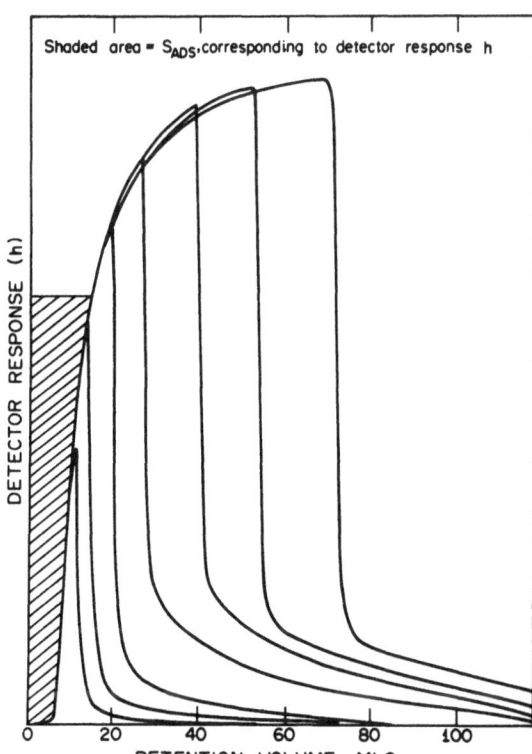

Fig. 12. Peak shapes for *n*-decane on poly(methyl methacrylate) beads at 25 °C. Injection size: 1.0, 0.7, 0.5, 0.3, 0.15, 0.06, 0.03 μl

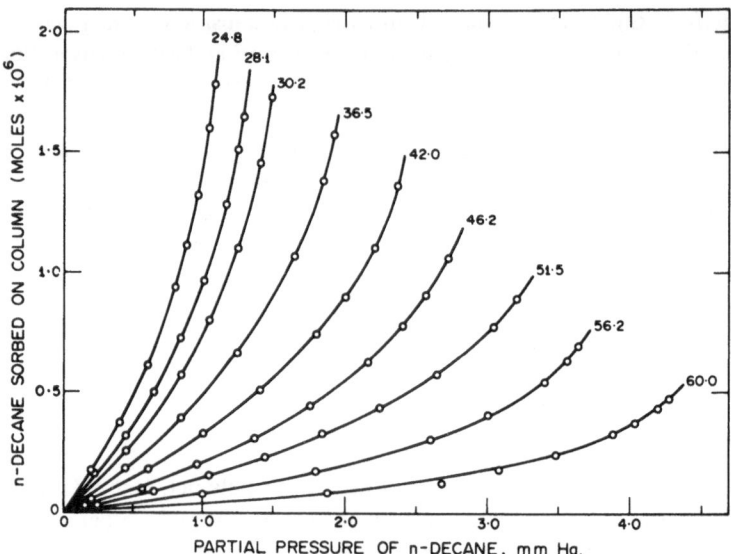

Fig. 13. Isotherms for n-decane on poly(methyl methacrylate) beads at several temperatures

represents the corresponding adsorption isotherms for decane on poly(methyl methacrylate) determined at several temperatures (*92*). It was found that these isotherms, identified as type III in Brunauer's classification, became linear at lower partial pressures as evidenced by sample size independent retention volumes and symmetrical elution peaks.

From the experimental adsorption isotherms both heats of adsorption and surface areas can be derived. For nonlinear adsorption isotherms the heat of adsorption varies with surface coverage and results are expressed as an isosteric heat of adsorption, at a specified coverage, a,

$$(-\Delta H_s)_a = -R\left(\frac{\partial \ln p_1}{\partial(1/T)}\right)_a \tag{26}$$

where $-\Delta H_s$ is the heat of adsorption (equal to but of opposite sign to the heat of desorption) and p_1 the partial pressure of solute in the gas phase. The surface area of the stationary phase can be determined by applying a BET approach to the experimental isotherm. For a two-parameter BET equation, one has

$$\frac{p_1/p_1^0}{v(1-p_1/p_1^0)} = 1/v_m C + (C - 1/v_m C)(p_1/p_1^0) \tag{27}$$

where p_1^0 is the saturated vapor pressure of the solute, v the volume of the solute on the surface, v_m the volume of the solute on the surface corresponding to monolayer formation and C a constant. It was found (*92*) that the surface areas determined from the adsorption isotherms through Eq. (27) were in good agreement with the geometric surface areas of the poly(methyl methacrylate) beads as well as for polymers deposited onto spherical glass beads.

More recently Mohlin and Gray (*93*) determined adsorption isotherms on cellulose fibers for a variety of adsorbates (solutes). From the experimental type II isotherms specific surface areas of the fibers were computed, for each solute, with the results given in Table 11. The agreement observed between the different solutes is quite remarkable considering that the area of the solute molecule on the polymer surface must be known or estimated. The surface area determined by nitrogen ad-sorption measurements at $-196°$ was included for the purpose of comparison. The slight disparity could possibly indicate that the area available to the smaller nitrogen molecule may be somewhat larger (1.9 compared to 1.6 $m^2 g^{-1}$).

Table 11. BET surface areas of cellulose fibers (*93*)

Solute	Temp., °C	C_{BET}	Area/ molecule, A^2	Surface area mol/g 10^6	m^2/g
n-Octane	25.8	5.76	45.8	5.93	1.63
n-Decane	25.8	6.65	51.6	5.22	1.62
	35.5	6.34	51.8	5.06	1.58
	45.0	5.80	51.9	4.97	1.55
	56.2	5.86	52.1	5.06	1.59
n-Dodecane	56.1	6.40	58.4	4.52	1.59
Dioxane	25.7	8.55	29.8	9.21	1.65
	40.2	7.26	30.1	9.43	1.71
Butanol	25.7	14.87	31.2	8.62	1.62
	49.8	14.58	31.7	8.47	1.61
Nitrogen	-196	–	16.2	–	1.9

Stoeckli and collaborators (*94–96*) combined gas chromatography and static methods in their studies of the surface properties of poly(vinyl chloride), poly(vinyl-idene fluoride) and poly(vinylidene chloride). Adsorption isotherms obtained by both methods of measurement were found in good agreement. In the case of poly(vinyl chloride) it was noted (*95*) that the surface properties depended on the thermal history of the sample. The specific surface area of the powdered polymer and the corresponding heat of adsorption changed when the polymer was heated for the first time through its T_g. Subsequent cycling through T_g did not reveal any further variations. This observation is to be correlated with a change in density of the material when heated above T_g for the first time. It was found (*94–96*) that the halogenated polymers investigated had similar, low energy surfaces. Only poly(vinyl chloride) exhibited the described modifications when heated above T_g.

Recently Tremaine and Gray (*97*) used the peak maximum elution method to determine adsorption isotherms of cellulose surfaces. In this method (*98*), the ad-sorption isotherm is obtained from the variation of the peak maximum retention volume with the solute sample size injected. Approximately 25 peaks are recorded for one such determination. Surface areas, heats and entropies of adsorption were computed from the experimental adsorption isotherms and the results were discussed in terms of theoretical models. It was found that, for both strongly and weakly interacting solutes, surface adsorption was the predominant retention mechanism. It was further shown (99) by these authors that BET surface areas can be obtained without first determining the adsorption isotherm. The peak maximum retention volumes are now directly fitted to a modified BET equation [Eq. (27)] to yield the monolayer capacity. It was found that surface areas so-determined were in agreement, to within 5%, with those derived from the adsorption isotherms while requiring less extensive data. This route has the further advantage of not requiring measurements at very low partial pressure where accuracy is rather poor.

At infinite dilution of the solute, adsorption isotherms usually become linear and the peak maximum retention volume is a characteristic of the polymer-solute system, with

$$V_N = K_a A \tag{28}$$

where V_N is the net retention volume, K_a the surface partition coefficient and A the surface area of the stationary phase. From the corresponding linear retention diagram the heat of adsorption can be computed, as outlined in Eq. (2). The partial molar heat of adsorption, Δh_a^∞, is obtained from the relation

$$\Delta h_a^\infty = R\partial \left[-\ln p_1^0 V_g - (p_1^0/RT)(B_{11} - V_1)\right]/\partial(1/T) \tag{29}$$

where all variables are as defined in Eq. (5). In general $\Delta . h_a^\infty$ is small and the heat of adsorption is very nearly equal to the heat of vaporization of the solute.

Equation (28) also makes possible a very easy determination of the surface area of any polymeric surface. Once the value of K_a has been determined for a certain polymer-solute system at a particular temperature, the surface area of any other finely divided sample of similar composition can be determined by measuring a retention volume with the same solute. Data on the partition coefficient for polystyrene with hexadecane have been reported by Braun and Guillet (13).

DeVries, Smit and Smith (100) reported heats of adsorption on cellulose triacetate (44.8% acetyl) and diacetate (25% acetyl) using alcohols and aromatics as solutes. They observed a systematic difference in heats of adsorption for each solute between both polymer stationary phases, indicating a difference in partial molar heats. They discussed their results in terms of nonspecific and hydrogen bonding interactions between the polymers and the solutes.

Sund et al. (101) and also Wallace et al. (20) reported retention characteristics of poly(vinyl chloride) powders of differing specific surface areas. It was shown that, once a calibration curve was established, gas chromatography could provide a rapid determination of surface areas of poly(vinyl chloride) powders. At temperatures below T_g the measured retention volume should be proportional to the surface partition coefficient and to the surface area of the stationary phase as expressed in Eq. (28). From the known surface areas of several samples the partition coefficient can be obtained, in turn allowing for the determination of the surface area of any sample from the measured retention volume. A similar correlation between BET surface areas and retention volumes was reported by Salovey et al. (102) for poly(vinyl chloride) powders.

Kiselev *et al.* (*103*) correlated the magnitude of the retention volume with the degree of coverage of a surface of rutile. Layers of poly(ethylene glycol) were deposited on surfaces of rutile by adsorption from solutions in toluene. It was found that the retention volume on rutile-poly(ethylene glycol) stationary phases decreased steadily with increasing surface coverage of poly(ethylene glycol), up to monolayer formation where no further decrease was detected. When rutile was replaced by graphitized carbon black, the amount of poly(ethylene glycol) required for complete coverage was considerably reduced, presumably due to the formation of thinner polymer films.

Braun and Guillet (*13*) recently studied the variation in surface area of polystyrene stationary phases coated on a diatomaceous support (Chromosorb G) as a function of coating thickness. It was found that the retention volume per gram of inert support remained unaffected over three orders of magnitude of film thickness. By comparison with a system of known surface area it was shown that only a fraction of the total area of the support accessible to nitrogen molecules ($\sim 1 \text{ m}^2\text{g}^{-1}$) was accessible to the larger polystyrene molecules ($0.1 \text{ m}^2\text{g}^{-1}$). As a result of the smaller surface area, it was indicated that film thickness of packed columns would be considerably larger than generally believed.

VII. Determination of Diffusion Coefficients

From the shape of the eluting solute peak, information can also be gained on the kinetic processes operative in a GC column. As the solute progresses from inlet to outlet, the band of solute molecules broadens due to diffusional spreading in both gas and liquid phases. Under suitable experimental conditions the diffusion coefficient of the solute in the polymer stationary phase can be determined from the width of a symmetrical eluted peak.

Van Deemter (*104*) first related peak broadening and column properties by an equation of the form

$$H = A + B/u + Cu \tag{30}$$

where H is the height-equivalent to the theoretical plate, u the linear velocity of carrier gas in the column, A, B, and C being constants independent of the velocity. The empirical constant A takes into account instrumental broadening while B is related to spreading in the gas phase. The constant C, pertaining to the finite time required to achieve equilibrium in the stationary phase, is given by,

$$C = (8/\pi^2)(d^2/D_l) [k/(1 + k)^2] \tag{31}$$

where d is the thickness of the stationary phase, D_l the diffusion coefficient of the solute in the polymer and k the partition ratio. k is equal to $(t_R - t_M)/t_M$ where t_R and t_M are the retention times for the solute and the noninteracting marker. In a typical experiment the plate heights, H, would be determined from the widths of the eluting solute peaks for several carrier gas velocities. The constant C would be obtained from the slope of the resulting Van Deemter plot (H versus u) which becomes linear at high velocities ($B/u \to 0$). From the known values of d and k the diffusion coefficient is readily evaluated.

Depending on the column configurations (packed, capillary, etc.) several formulations of Eq. (30) have been suggested. In the present case column parameters must be designed so as to magnify the effects of slow diffusion in the stationary phase. This is quite easily achieved with polymer stationary phases since their diffusion coefficients are usually smaller by two orders of magnitude than those of low molecular weight liquids. It should also be noted that measurements must be performed under equilibrium conditions, $i.e.$, at temperatures in excess of $T_g + 50°$.

Gray and Guillet (105) studied the diffusion of low molecular weight solutes in polymer stationary phases. It was found that among all column configurations (packed, open columns, etc.) polymers coated on spherical glass beads represented an optimum for their ease of manufacture and definition of film thickness. A summary of their results for the diffusion coefficients of hydrocarbons in polyethylene is given in Table 12. It was noted that at temperatures below the melting of the stationary phase the activation energy for diffusion decreased with increasing temperature. Above melting linear Arrhenius plots were obtained. Comparison of their results with available literature data was most satisfactory.

Table 12. Diffusion coefficients from gas chromatographic measurements on low density polyethylene (105)

Probe	Temp., °C	Van Deemter C term sec x 10^3	Partition ratio, k	$k/(1 + k)^2$	Diffusion coefficients, D cm^{2}sec^{-1} x 10^8
n-Tetradecane	125	20 ± 7	20.4	0.045	0.85
	140	13.0 ± 0.06	11.3	0.075	2.2
	150	10.3 ± 0.4	7.83	0.100	3.7
	160	12.5 ± 0.6	5.47	0.131	4.1
	170	8.6 ± 0.5	3.88	0.163	7.4
n-Decane	30	30.0 ± 1.8	34.3	0.028	0.35
	50	24.6 ± 0.7	13.2	0.066	1.03
	60	31.8 ± 0.7	10.0	0.083	1.00
	65	29.9 ± 1.5	8.0	0.099	1.28
	80	38.2 ± 1.0	5.3	0.133	1.34
Benzene	25	114 ± 6	1.50	0.240	0.82
BHA (106)	137	–	44.2	–	7.38
BHT (106)	137	–	57.0	–	4.70

More recently Braun et al. (106) investigated the diffusion of stabilizers in molten polyethylene. Their results for butylated hydroxy toluene (BHT), butylated hydroxy anisole (BHA) and several hydrocarbons are given in Fig. 14. The Arrhenius

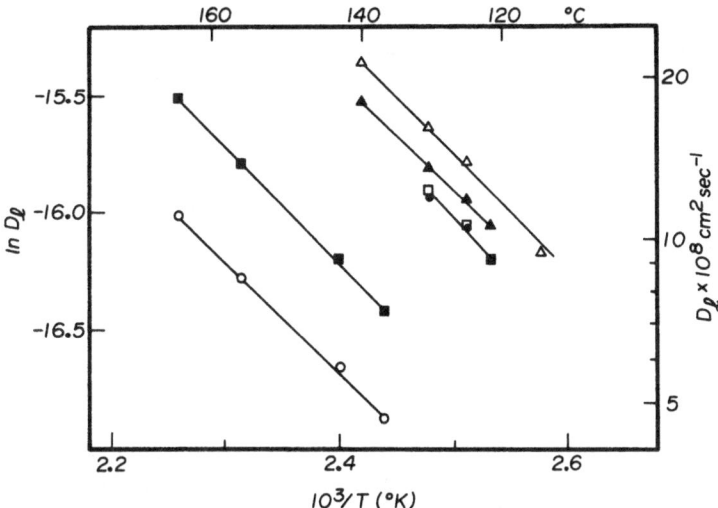

Fig. 14. Arrhenius plots of diffusion coefficients in low density polyethylene: (△) tetralin, (▲) *trans*-decalin, (□) *cis*-decalin, (●) *n*-dodecane, (■) BHA, (○) BHT

plots were linear for the temperature range investigated, with activation energies of 10–12 kcal/mol.

It is worth noting that the GC method is particularly suitable for molten polymers since there is no need for the polymer to be self-supporting. Moreover, the diffusion coefficients so-determined should effectively correspond to infinite dilution of solute in the polymer, without recourse to extrapolation procedures.

VIII. Other Applications

Wallace *et al.* (*20*) correlated GC retention volumes of several poly(vinyl chloride) powders with their uptake of plasticizer. Since the diffusion of plasticizers into polymer powders is controlled by the external surface area, the diffusion coefficient of the plasticizer and some shape factor, a correlation with GC measurements could be expected. It was found that plasticizer absorption ("drying") took place only when the polymer was heated to a temperature immediately above the glass transition temperature as defined by the minimum of the experimental retention diagram.

Perrault *et al.* (*22, 107*) investigated the curing of polymeric binders by gas chromatography. It was found that both filled and non-filled stationary phases could be studied, thereby approximating quite closely the actual experimental conditions. The progress of the curing reaction was followed by recording the variation of retention volume with time. The effects of several commonly used fillers were also assessed.

Niederstrass (*108*) used gas chromatography to characterize polymer stationary phases of differing styrene/butadiene contents. These included mechanical mixtures of polystyrene and polybutadiene, statistical and graft copolymers as well as block copolymers of styrene and butadiene. It was found that the composition of the stationary phase could be determined from the measured retention volume, using a calibration curve. However, it did not prove possible to discriminate between mixtures of polymers and graft or block copolymers of identical overall composition.

IX. Conclusions

As a result of the experimental work summarized in this review the value of the gas chromatographic method for studying polymers seems to be well established. Surface and bulk properties of polymers can be measured both from a thermodynamic and kinetic point of view. Because of its simplicity and precision it should become the method of choice for the study of thermodynamic interactions of small molecule "probes" or solutes in systems where the polymer is the major phase. Further advances in the kinetic theory of the GC process should provide even more reliable data about time dependent processes such as diffusion, adsorption, complex formation and possibly even chemical reaction.

The use of GC to determine polymer crystallinity is of particular interest, since it is one of the few methods which do not depend on X-ray determinations of crystal structure for calibration. Presumably the method will be used more generally when suitable instruments are commercially available.

Note Added in Proof

Since this manuscript was sent to the publisher, several papers on inverse gas chromatography appeared in the technical literature, which ought to be included in the present review. Hsiung and Cates (*109*) investigated the GC behavior of different poly(ethylene terephthalate)s with several organic solutes in the vicinity of the glass transition temperature. Maloney and Prausnitz measured the solubility of ethylene in low density polyethylene both at high temperatures (*110*) and at industrial-separation pressures (*111*). Stiel and Harnish (*112*) reported the solubilities of gases and liquids in polystyrene at pressures up to 4 atm. Sunol and Barker (*113*) determined the activity coefficients of pinenes in poly(propylene sebacate) at finite concentration of solute by the technique of elution on a plateau. An exhaustive compilation of polymer-solute interaction parameters obtained both by GLC and by static methods was published by Bonner (*114*). Martire (*115*) commented on the microscopic partition theory recently proposed by Purnell and collaborators (*80*). Kong and Hawkes (*116*) used GLC to determine diffusion coefficients of small organic solutes in commercial, uncrosslinked silicones of varying molecular weights.

X. References

1. Smidsrod, O., and Guillet, J. E.: Macromolecules **2**, 272 (1969).
2. Purnell, J. H.: Gas chromatography. New York: Wiley 1962.
3. Littlewood, A. B.: Gas chromatography. New York: Academic Press 1970.
4. Kiselev, A. V., and Yashin, Ya. I.: Gas adsorption chromatography. New York: Plenum 1969.
5. Young, C. L.: Chromatogr. Rev. **10**, 129 (1968).
6. Alishoev, V. R., Berezkin, V. G., and Mel'nikova, Yu. V.: Russ. J. Phys. Chem. **39**, 105 (1965).
7. Guillet, J. E., and Stein, A. N.: Macromolecules **3**, 102 (1970).
8. Guillet, J. E. in: New developments in gas chromatography (Ed. J. H. Purnell). New York: Wiley 1973.
9. Guillet, J. E.: J. Macromol. Sci. – Chem. **A4**, 1669 (1970).
10. Lavoie, A., and Guillet, J. E.: Macromolecules **2**, 443 (1969).
11. Galin, M., and Guillet, J. E.: J. Polym. Sci., Polym. Lett. Ed. **11**, 233 (1973).
12. Braun, J.-M., Lavoie, A., and Guillet, J. E.: Macromolecules **8**, 311 (1975).
13. Braun, J.-M., and Guillet, J. E.: Macromolecules **8**, 882 (1975).
14. Braun, J.-M.: Ph. D. Thesis, University of Toronto, 1975.
15. Galassi, S., and Audisio, G.: Makromol. Chem. **175**, 2975 (1974).
16. Braun, J.-M., and Guillet, J. E.: J. Polym. Sci., Polym. Chem. Ed., in press.
17. Liebman, S. A., Ahlstrom, D. H., and Foltz, C. R.: J. Chromatogr. **67**, 153 (1972).
18. Yamamoto, Y., Tsuge, S., and Takeuchi, T.: Bull. Chem. Soc. Japan **44**, 1145 (1971).
19. Gray, D. G., and Guillet, J. E.: J. Polym. Sci., Polym. Lett. Ed.**12**, 831 (1974).
20. Wallace, J. R., Kozak, P. J., and Noel, F.: SPE J. **26**, 43 (1970).
21. Valentin, N., Chabert, B., Chauchard, J., and Edel, G.: Bull. Scient. ITF **3**, 5 (1974).
22. Perrault, G., Tremblay, M., Bedard, M., Duchesne, G., and Voyzelle, R.: Eur. Polym. J. **10**, 143 (1974).
23. Ateya, K., Chabert, B., Chauchard, J., and Edel, G.: C. R. Acad. Sci. Paris **274C**, 506 (1972).
24. Chabert, B., Chauchard, J., Edel, G., and Soulier, J. P.: Eur. Polym. J. **9**, 993 (1973).
25. Chabert, B., Chauchard, J., Edel, G., Soulier, J. P., and Valentin, N.: J. Chim. Phys., Phys.-Chim. Biol. **72**, 215 (1975).
26. Nakamura, S., Shindo, S., and Matsuzaki, K.: J. Polym. Sci., B **9**, 591 (1971).
27. Calugaru, E.-M., and Schneider, I. A.: Eur. Polym. J. **10**, 729 (1974).
28. Martin, R. L.: Anal. Chem. **33**, 347 (1961).
29. Braun, J.-M., and Guillet, J. E.: Macromolecules **9**, 340 (1976).
30. Nesterov, A. Y., and Lipatov, Y. S.: Vysokomol Soyed. **A15**, 2601 (1973); Macromolecules, **8**, 889 (1975).
31. Martire, D. E. in: Progress in gas chromatography (Ed. J. H. Purnell). New York: Wiley 1968.
32. Gray, D. G., and Guillet, J. E.: Macromolecules **7**, 244 (1974).
33. Courval, G. J., and Gray, D. G.: Macromolecules **8**, 916 (1975).
34. Braun, J.-M., and Guillet, J. E.: Macromolecules **8**, 557 (1975).
35. Braun, J.-M., and Guillet, J. E.: Macromolecules, in press.
36. Fujita, H., Kishimoto, A., and Matsumoto, K.: Trans. Faraday Soc. **56**, 424 (1960).
37. Cruikshank, A. J. B., Windsor, M. L., and Young, C. L.: Proc. Roy. Soc. **A295**, 259, 271 (1966).
38. Conder, J. R., and Purnell, J. H.: Trans. Faraday Soc. **64**, 1505 (1968).
39. Patterson, D., Tewari, Y. B., Schreiber, H. P., and Guillet, J. E.: Macromolecules **4**,356 (1971).
40. Roberts, G. L., and Hawkes, S. J.: J. Chromatogr. Sci. **11**, 16 (1973).
41. Delmas, G., Patterson, D., and Somcynsky, T.: J. Polym. Sci. **57**, 79 (1962); Patterson, D.: J. Polym. Sci., C **16**, 3379 (1968).
42. Flory, P. J., Orwoll, R. A., and Vrij, A.: J. Am. Chem. Soc. **86**, 3507, 3515 (1964); Orwoll, R. A., and Flory, P. J.: *ibid.* **89**, 6814, 6822 (1967); Eichinger, B. E., and Flory, P. J.: Trans. Faraday Soc. **64**, 2035 (1968).

43. Prigogine, I., (with A. Mathot and A. Bellemans): The molecular theory of solutions. Amsterdam, North Holland Publishing Co. 1957.
44. Summers, W. R., Tewari, Y. B., and Schreiber, H. P.: Macromolecules 5, 12 (1972).
45. Tewari, Y. B., and Schreiber, H. P.: Macromolecules 5, 329 (1972).
46. Chahal, R. S., Kao, W. P., and Patterson, D.: JCS Faraday I 69, 1834 (1973).
47. Morimoto, S.: Makromol. Chem. 133, 197 (1970).
48. Hammers, W. E., and DeLigny, C. L.: Rec. Trav. Chim. 90, 912 (1971).
49. Hammers, W. E., and DeLigny, C. L.: J. Polym. Sci., C 39, 273 (1972).
50. Marcille, P., Audebert, R., and Quivoron, C.: C. R. Acad. Sci. Paris, C 277, 9 (1973); J. Chim. Phys. 1, 78 (1975).
51. Newman, R. D., and Prausnitz, J. M.: AIChE J. 19, 704 (1973); J. Phys. Chem. 76, 1492 (1972).
52. Leung, Y. K., and Eichinger, B. E.: J. Phys. Chem. 78, 60 (1974); Macromolecules 7, 685 (1974).
53. Schreiber, H. P., Tewari, Y. B., and Patterson, D.: J. Polym. Sci., Polym. Phys. Ed. 11, 15 (1973).
54. Patterson, D., Tewari, Y. B., and Schreiber, H. P.: JCS Faraday II 68, 885 (1972).
55. Lichtenthaler, R. N., Liu, D. D., and Prausnitz, J. M.: Ber. Buns. Gesellsch. 78, 470 (1974).
56. Lichtenthaler, R. N., Abrams, D. S., and Prausnitz, J. M.: Can. J. Chem. 51, 3071 (1973).
57. Hammers, W. E., and DeLigny, C. L.: J. Polym. Sci., Polym. Phys. Ed. 12, 2065 (1974).
58. Hammers, W. E., Bos, B. C., Loomans, Y. J. W. A., and DeLigny, C. L.: J. Polym. Sci., Polym. Phys. Ed. 13, 401 (1975).
59. Covitz, F. H., and King, J. W.: J. Polym. Sci., A1 10, 689 (1972).
60. Varsano, J. L., and Gilbert, S. G.: J. Pharm. Sci. 62, 87 92 (1973).
61. Gosh, S. K.: Makromol. Chem. 143, 181 (1971).
62. Newman, R. D., and Prausnitz, J. M.: AIChE J. 19, 704 (1973).
63. Cheng, Y. L., and Bonner, D. C.: Macromolecules 7, 687 (1974); Chang, Y. H., and Bonner, D. C.: J. Appl. Polym. Sci. 19, 2439 (1975).
64. Lichtenthaler, R. N., Newman, R. D., and Prausnitz, J. M.: Macromolecules 6, 650 (1973).
65. Lichtenthaler, R. N., Prausnitz, J. M., Su, C. S., Schreiber, H. P., and Patterson, D.: Macromolecules 7, 136 (1974).
66. Lichtenthaler, R. N., Liu, D. D., and Prausnitz, J. M.: Macromolecules 7, 565 (1974).
67. Reichert, K. H.: J. Oil Col. Chem. Assoc. 54, 887 (1971).
68. Newman, R. D., and Prausnitz, J. M.: J. Paint Techn. 45, no. 585, 33 (1973).
69. McCathern, V. R., and Thompson, D. C.: Rubber World 167, 33 (1972).
70. Conder, J. R., and Purnell, J. H.: Trans. Faraday Soc. 64, 3100 (1968); 65, 824, 839 (1969).
71. Brockmeier, N. F., McCoy, R. W., and Meyer, J. A.: Macromolecules, 5, 130, 464 (1972); 6, 176 (1973).
72. Brockmeier, N. F., Carlson, R. E., and McCoy, R. W.: AIChE J. 19, 1133 (1973).
73. Chang, Y. H., and Bonner, D. C.: J. Appl. Polym. Sci. 19, 2457 (1975).
74. Deshpande, D. D., Patterson, D., Schreiber, H. P., and Su, C. S.: Macromolecules 7, 530 (1974).
75. Patterson, D., and Schreiber, H. P.: Soc. Plast. Eng., Tech. Pap. 21, 120 (1975).
76. Su, C. S., Patterson, D., and Schreiber, H. P.: J. Appl. Polym. Sci., 20, 1025 (1976).
77. Nesterov, A. E., and Lipatov, Y. S.: Vysokomol. Soyed. A16, 1919 (1974).
78. Nesterov, A. E., and Lipatov, Y. S.: Vysokomol. Soyed. A17, 671 (1975).
79. Olabisi, O.: Macromolecules 8, 316 (1975).
80. Purnell, J. H., and Vargas de Andrade, J. M.: J. Am. Chem. Soc. 97, 3585, 3590 (1975); Laub, R. J., and Purnell, J. H.: ibid., 98, 30, 35 (1976).
81. McCrea, P. F. in: New Developments in gas chromatography (Ed. J. H. Purnell). New York: Wiley 1973.
82. Gray, D. G., and Guillet, J. E.: Macromolecules 4, 129 (1971).
83. Stein, A. N., Gray, D. G., and Guillet, J. E.: Br. Polym. J. 3, 175 (1971).
84. Braun, J.-M., and Guillet, J. E.: J. Polym. Sci., Polym. Chem. Ed. 13, 1119 (1975).
85. Courval, G., and Gray, D. G.: Macromolecules 8, 326 (1975).

86. Nesterov, A. E., and Lipatov, Y. S.: Sint. Fiz. – Khim. Polim. 15, 60 (1975).
87. Braun, J.-M., and Guillet, J. E.: manuscript in preparation.
88. Hollis, O. L.: Anal. Chem. 38, 309 (1966).
89. Gearhart, H. L., and Burke, M. F.: J. Chromatogr. Sci. 11, 411 (1973); Rakshieva, N. R., Novak, J., Vicar, S., and Janak, J.: J. Chromatogr. 91, 51 (1974).
90. Gluekauf, E.: J. Chem. Soc., 1302 (1947).
91. Edel, G., and Chabert, B.: C. R. Acad. Sci. Paris C268, 226 (1969); ibid. C267, 54 (1968).
92. Gray, D. G., and Guillet, J. E.: Macromolecules 5, 316 (1972).
93. Mohlin, U. B., and Gray, D. G.: J. Coll. Interf. Sci. 47, 747 (1974).
94. Houriet, J. P., Ghiste, P., and Stoeckli, F.: Helv. Chim. Acta 57, 851 (1974).
95. Janneret, C., and Stoeckli, H. F.: ibid., 56, 2509 (1973).
96. Stoeckli, H. F.: ibid., 55, 101 (1972).
97. Tremaine, P. R., and Gray, D. G.: J. Chem. Soc., Faraday I 71, 2170 (1975).
98. Huber, J. F. K., and Gerritse, R. G.: J. Chromatogr. 58, 137 (1971).
99. Tremaine, P. R., and Gray, D. G.: Anal. Chem. 48, 380 (1976).
100. DeVries, M. J., and Smit, J. H.: J. S. Afr. Chem. Inst. 20, 11 (1967); Smit, J. H., Smith, J. H., and DeVries, M. J.: ibid., 20, 144 (1967).
101. Sund, E., Haanaes, E., Smidsrφd, O., and Ugelstad, J.: J. Appl. Polym. Sci. 16, 1869 (1972).
102. Salovey, R., Cortellucci, R., and Roaldi, A.: Polym. Eng. Sci. 14, 120 (1974).
103. Kiselev, A. V., Kovaleva, N. V., Khopina, V. V., and El'tekov, Y. A.: Vyskomol. Soyed. A16, 1142 (1974).
104. Van Deemter, J. J., Zuiderweg, F. J., and Klinkenberg, A.: Chem. Eng. Sci. 5, 271 (1956).
105. Gray, D. G., and Guillet, J. E.: Macromolecules 6, 223 (1973).
106. Braun, J.-M., Poos, S., and Guillet, J. E.: J. Polym. Sci., Polym. Lett. Ed., in press.
107. Perrault, G., Duchesne, G., and Tremblay, M.: Eur. Polym. J. 10, 747 (1974).
108. H. Niederstrass, H., Dr. rer. nat., Braunschweig, 1975.
109. Hsiung, P. L., and Cates, D. M.: J. Appl. Polym. Sci. 19, 3051 (1975).
110. Maloney, D. P., and Prausnitz, J. M.: AIChE J. 22, 74 (1976).
111. Maloney, D. P., and Prausnitz, J. M.: Ind. Eng. Chem., Process Res. Dev. 15, 216 (1976).
112. Stiel, L. I., and Harnish, D. F.: AIChE J. 22, 117 (1976).
113. Sunol, A. B., and Barker, P. E.: J. Chromatogr. Sci. 13, 541 (1975).
114. Bonner, D. C.: J. Macromol. Sci., Rev. Chem. C13, 263 (1975).
115. Martire, D. E.: Anal. Chem. 48, 398 (1976).
116. Kong, J. M., and Hawkes, S. J.: Macromolecules 8, 685 (1975).

Received December 18, 1975.

Author-Index Volume 1–21

B. Vollmert

POLYMER CHEMISTRY

Translated from the German by E. H. Immergut,
New York
With 630 figures. XVII, 652 pages. 1973

This book gives a comprehensive coverage of the synthesis of polymers and their reactions, structure, and properties. The treatment of the reactions used in the preparation of macromolecules and in their transformation into cross-linked materials is particularly detailed and complete. The book also gives an up-to-date presentation of other important topics, such as enzymatic and protein synthesis, solution properties of macromolecules, polymer crystallization, and properties of polymers in the solid state.
The content and presentation of Professor Vollmert's book is more encompassing than most existing treatises, and its numerous figures and tables convey a wealth of data, never, however, at the expense of intellectual clarity or educational value.
The presentation is mainly on a fundamental and general level and yet the reader—student or professional—is gradually and almost casually introduced to all important natural and synthetic polymers. Complicated phenomena are explained with the aid of the simplest available examples and models in order to ensure complete understanding. However, the reader is also encouraged to think for himself and even to criticize the author's point of view.
All of the chapters have been revised and enlarged from the German edition, and many of the sections are entirely new.

Contents
Introduction. — Structural Principles. — Synthesis and Reactions of Macromolecular Compounds. — The Properties of the Individual Macromolecule. — States of Macromolecular Aggregation.

Springer-Verlag
Berlin Heidelberg New York

Chemie, Physik und Technologie der Kunststoffe/ Properties and Applications of Plastics

Springer-Verlag
Berlin Heidelberg New York